The Many Faces of Complexity

The Many Faces of Complexity
An Interdisciplinary Approach to Beauty

Aleksandar I. Zečević

Dept. of Electrical Engineering
Santa Clara University

www.scriptaaesthetica.com

Copyright © 2018 – Aleksandar I. Zecevic. All rights reserved. No part of this publication may be reprinted, reproduced, transmitted or utilized in any form or by any electronic, mechanical, or other means now known or hereafter invented, including photocopying, microfilming and recording, or in any information retrieval system without the written permission of the author.

ISBN: 978-1987652963

Cover illustration: *Julia Set*

Contents

Preface   v

I   Information   1

1   Classical Information Theory   3
- 1.1   Encoding and Error Probabilities 3
  - 1.1.1   The Huffman Code 7
- 1.2   Information Entropy 10
  - 1.2.1   Joint Entropy 14
  - 1.2.2   Conditional Entropy 16
- 1.3   Interpretations of Information Entropy 20
  - 1.3.1   Entropy and Encoding 20
  - 1.3.2   Entropy and Probable Sequences 27

2   Information in Biology   35
- 2.1   Protein Synthesis 35
- 2.2   Information Processing in Biological Systems 38
- 2.3   Redundancy and "Genetic Noise" 41

3   Quantum Information   47
- 3.1   An Overview of Quantum Mechanics 49
- 3.2   Quantum Computing 53
  - 3.2.1   Single Qubit Gates 55
  - 3.2.2   The C-NOT Gate 58
  - 3.2.3   Classical Computations on Quantum Computers 65
  - 3.2.4   Quantum Algorithms 69

4   Notes and References   75
- 4.1   Notes for Chapters 1-3 75
- 4.2   Further Reading 77

II Self-Similarity and Fractals — 79

5 Self-Similarity — 81
- 5.1 Power Laws and Spectral Density — 81
- 5.2 Self-Similarity in the Time Domain — 88

6 Fractals — 95
- 6.1 Fractal Dimensions — 95
- 6.2 Multifractals — 109
- 6.3 Constructing Fractals of a Given Dimension — 115
- 6.4 Fractals as Attractors — 120
- 6.5 Generalized Dimensions — 128

7 Notes and References — 135
- 7.1 Notes for Chapters 5-6 — 135
- 7.2 Further Reading — 136

III Complexity and Self-Organization — 139

8 Nonlinearity, Chaos and Catastrophes — 141
- 8.1 Nonlinear Systems — 141
  - 8.1.1 Types of Attractors in Nonlinear Systems — 146
  - 8.1.2 Hypersensitivity — 153
- 8.2 Catastrophes — 163

9 Cellular Automata — 175
- 9.1 Properties of Cellular Automata — 175
- 9.2 An Analytic Classification of Rules — 190
- 9.3 Random Boolean Networks — 197

10 Self-Organized Criticality — 205
- 10.1 Sand Piles and Autocatalysis — 205
  - 10.1.1 Sand Piles — 205
  - 10.1.2 Autocatalysis — 211
- 10.2 Self-Organization and Information — 215
  - 10.2.1 The Maximum Information Principle — 220
  - 10.2.2 A General Framework — 226

11 Notes and References — 229
- 11.1 Notes for Chapters 8-10 — 229
- 11.2 Further Reading — 231

CONTENTS

IV Mathematical Infinity — 235

12 Numbers, Limits and Infinite Sums — 237
- 12.1 Integers, Primes and Real Numbers 237
  - 12.1.1 Prime Numbers 237
  - 12.1.2 Real Numbers 239
- 12.2 Division by Zero 246
- 12.3 Infinite Sums 249

13 Introduction to Infinite Sets — 255
- 13.1 What is a Set? 255
  - 13.1.1 The Intuitive Approach 255
  - 13.1.2 The Axiomatic Approach 258
  - 13.1.3 Pure Sets 260
- 13.2 Ordinals 262
  - 13.2.1 Infinite Ordinals 263
- 13.3 Cardinals 267
  - 13.3.1 Hierarchies of Infinities 270
  - 13.3.2 Transfinite Arithmetic with Cardinals 274
- 13.4 Infinite Ordinals and Cardinals 276
- 13.5 Infinitesimals and Hyperreals 277

14 Notes and References — 283
- 14.1 Notes for Chapters 12-13 283
- 14.2 Further Reading 286

Preface

This book is the final part of a trilogy which explores the role of beauty in art, science and mathematics. Unlike the other two volumes, this one is purely technical and requires a solid background in differential equations, as well as a basic knowledge of probability theory. Because of that, it is suitable mainly for individuals who are pursuing (or already have) degrees in science and engineering.

Although its primary purpose is to serve as a textbook for my course on interdisciplinary aesthetics at Santa Clara University, I believe that this book will also be of interest to anyone who would like to broaden their technical expertise and explore certain fields that they may not be familiar with. It has a lot to offer in that respect, since it covers a wide variety of topics (ranging from information theory and quantum computing to fractal geometry, cellular automata and nonlinear dynamics).

While some of these topics appear to have little in common, it is important to keep in mind that they all share a connection with beauty. This connection is sometimes obvious and sometimes hidden, but can always be found if one chooses to look hard enough. I have made an effort to point out these links whenever possible, but have generally refrained from discussing them in greater detail. Those who would like to explore such questions more deeply are encouraged to read my other two books on the subject, *The Beauty of Nature and the Nature of Beauty*, and *Ten Dialogues About Art and Beauty*.

The topics presented in this book are divided into four more or less self-contained sections, which deal with information, fractals, complexity and infinite sets. Since each section contains both introductory and advanced material, it can be read on several different levels, depending on the reader's technical background. Having said that, however, I should add that anyone who studies the text carefully is bound to learn something new (and hopefully valuable). I think this is very important, since we live in an age that promotes narrow specialization and provides few incentives for exploring other disciplines. History teaches us that such an attitude can be counterproductive, since we are much more likely to make unexpected connections if our knowledge base is broad (the Renaissance has shown that very clearly). These connections needn't always be particularly deep or involve major discoveries, but anyone who has experienced such moments of sudden insight knows how profoundly rewarding they can be.

If the purpose of higher education is to facilitate this process and prepare our students for such experiences, then we ought to encourage them to cross disciplinary boundaries and expand their horizons in every possible direction. We must make sure, however, that they do so in a way that is systematic and disciplined, and avoids simplistic explanations. It is largely for this reason that I chose to present each of the topics with as much rigor as possible, and include suggestions for further reading for those who are interested in digging a bit deeper.

Acknowledgments

I would like to thank my son Stefan Zecevic for his help in creating the illustrations for this book. He produced all the figures, and performed the necessary simulations in Matlab, Mathematica, and R.

Part I

Information

Chapter 1

Classical Information Theory

1.1 Encoding and Error Probabilities

A typical schematic diagram for a communication system is shown in Fig 1.1.

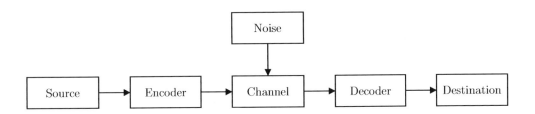

Figure 1.1: Components of a typical communication system.

A key step in conveying information from the source to the destination is the encoding process, which maps each message into an "object" that is suitable for transmission. This "object" could be a sequence of bits, an analog waveform, or even a biochemical pattern (such as the one that living organisms use to synthesize proteins).

Since communication channels are never ideal, there is always a chance that the original message could be corrupted by noise, and that it could be misinterpreted on the receiving end. This possibility is usually accounted for by associating an *error probability* with each signal that is emitted by the source. To see how this probability might be reduced, let us consider a simple example in which the source produces only zeros and ones, and the error probability is $p = 0.25$ in both cases. An obvious way to improve the reliability of the transmission would be to encode each emitted bit in the manner illustrated in Table 1.1.

Original Message		Encoding
0	→	000
1	→	111

Table 1.1. An encoding strategy based on redundancy.

Such a strategy would obviously decrease the likelihood of an error on the receiving end, since each message is sent multiple times. We must keep in mind, however, that this would also reduce the transmission rate by a factor of three, which is clearly undesirable. It is fair to say, therefore, that there is an inherent trade-off between the reliability of the messages and the speed with which they are sent.

To get a better sense for how this type of encoding reduces the error probability, let us consider how an incoming message might produce an erroneous reading. For the sake of simplicity, we will assume that the decoder uses a "majority selection" algorithm, which identifies the message with the bit that appears more than 50% of the time. Under such circumstances, there are four possible scenarios that could lead to an error in interpreting bit 1:

Received Signal	Interpretation
001	0
010	0
100	0
000	0

Table 1.2. Four signal patterns that produce an incorrect interpretation.

Recalling that the error probability for each transmitted bit is p, the likelihood of an erroneous interpretation when each bit is sent three times can be computed as

$$P(e) = 3p^2(1-p) + p^3 \tag{1.1}$$

The first term in this expression corresponds to the case when either 001, 010 or 100 are received, while the second one takes into account the possibility that all three received bits could be incorrect. Setting $p = 0.25$ in (1.1) produces $P(e) = 0.15625$, which is significantly smaller than the original error probability.

The above example clearly shows that the error probability can be reduced if we allow for a certain amount of redundancy in the way the message is transmitted. The following theorem generalizes this result, and examines what happens when each bit is encoded using $2n+1$ repetitions.

Theorem 1.1. Suppose that we use an encoding in which each bit is repeated $2n+1$ times. If the error probability β for each bit is less than 0.5,

1.1. ENCODING AND ERROR PROBABILITIES

then the expected number of erroneous bits in the sequence is *always* less than $n+1$.

Proof. We begin by observing that the probability of encountering k erroneous bits in a sequence of length $2n+1$ is

$$P(k) = \binom{2n+1}{k} \beta^k (1-\beta)^{2n+1-k} \qquad (1.2)$$

(the first term in (1.2) reflects the fact that there are multiple ways in which such a scenario can occur). Given this expression, we can calculate the expected number of incorrect bits as

$$E = \sum_{k=1}^{2n+1} k \cdot P(k) = \sum_{k=1}^{2n+1} k \cdot \binom{2n+1}{k} \beta^k (1-\beta)^{2n+1-k} \qquad (1.3)$$

In order to simplify this sum, we should recall that

$$k \cdot \binom{2n+1}{k} = \frac{k(2n+1)!}{k!(2n-k+1)!} = \frac{(2n+1)!}{(k-1)!(2n-k+1)!} \qquad (1.4)$$

Setting $m = k-1$, (1.3) becomes

$$\begin{aligned} E &= \sum_{m=0}^{2n} \frac{(2n+1)!}{m!(2n-m)!} \beta^{m+1}(1-\beta)^{2n-m} = \\ &= (2n+1)\beta \sum_{m=0}^{2n} \frac{(2n)!}{m!(2n-m)!} \beta^m (1-\beta)^{2n-m} \end{aligned} \qquad (1.5)$$

Using the fact that

$$\frac{(2n)!}{m!(2n-m)!} = \binom{2n}{m} \qquad (1.6)$$

and recalling the binomial expansion formula, we finally obtain

$$E = (2n+1)\beta \cdot [\beta + (1-\beta)]^{2n} = (2n+1)\beta \qquad (1.7)$$

Our initial assumption that $\beta < 0.5$ ensures that

$$E = (2n+1)\beta < n + \frac{1}{2} < n+1 \qquad (1.8)$$

for *any* choice of n, which implies that we can always expect the message to be decoded *correctly* (since the expected number of erroneous digits is less than half of the total). **Q.E.D.**

CHAPTER 1. CLASSICAL INFORMATION THEORY

The following theorem strengthens this result, and shows that the likelihood of an incorrect interpretation becomes vanishingly small as $n \to \infty$.

Theorem 1.2. Let S_{2n+1} denote the number of erroneous bits in a message that uses $2n + 1$ repetitions to encode each bit produced by the source. If the error probability for each bit is $\beta < 0.5$, the likelihood of an erroneous decoding approaches zero when $n \to \infty$.

Proof. We begin by observing that the probability of an erroneous decoding can be expressed as

$$p_n(e) = \Pr\left\{\frac{S_{2n+1}}{2n+1} \geq 0.5\right\} \tag{1.9}$$

To show that $p_n(e) \to 0$ as $n \to \infty$, it will be helpful to associate a random variable X_i with each of the $2n+1$ bits that are used to encode a single message. The simplest way to do this is to assume that $X_i = 1$ if the corresponding bit is erroneous, and $X_i = 0$ otherwise.

It is easy to see that the expected value of each X_i defined in this manner equals β. Indeed, since $\Pr(X_i = 1) = \beta$, we have that

$$E(X_i) = 1 \cdot \beta + 0 \cdot (1-\beta) = \beta \tag{1.10}$$

It is also obvious that S_{2n+1} can be expressed as

$$S_{2n+1} = \sum_{i=1}^{2n+1} X_i \tag{1.11}$$

which allows us to interpret the term $S_{2n+1}/2n+1$ as

$$\frac{S_{2n+1}}{2n+1} = \frac{1}{2n+1} \sum_{i=1}^{2n+1} X_i \tag{1.12}$$

This formulation is convenient because it allows us to establish the asymptotic behavior of $p_n(e)$ using the "weak law of large numbers".[1] This law concerns a collection of n random numbers $\{X_1, X_2, \ldots, X_n\}$ which have *identical* expectations $E(X_1) = E(X_2) = \ldots = E(X_n) = \mu$. It states that the average value

$$\bar{X}_n = \frac{1}{n} \sum_{i=1}^{n} X_i \tag{1.13}$$

of such a set must satisfy

$$\lim_{n \to \infty} \Pr\left\{|\bar{X}_n - \mu| \geq \varepsilon\right\} = 0 \tag{1.14}$$

1.1. ENCODING AND ERROR PROBABILITIES

for *any* $\varepsilon > 0$. Since (1.12) represents a variant of expression (1.13), it follows that

$$\lim_{n \to \infty} \Pr\left\{\left|\frac{S_{2n+1}}{2n+1} - \beta\right| \geq \varepsilon\right\} = 0 \qquad (1.15)$$

for any $\varepsilon > 0$. This is equivalent to saying that the ratio $S_{2n+1}/2n+1$ approaches β with probability equal to 1 when $n \to \infty$. Recalling that $\beta < 0.5$, we have that

$$\lim_{n \to \infty} \Pr\left\{\frac{S_{2n+1}}{2n+1} \geq 0.5\right\} = 0 \qquad (1.16)$$

which is what we set out to prove. **Q.E.D.**

Although Theorem 1.2 shows that increasing the value of n can minimize the likelihood of a decoding error, we should reiterate that such a strategy has the opposite effect on the transmission rate. This is an obvious consequence of the fact that the sequence of bits associated with each generated message becomes longer. It turns out, however, that it is possible to achieve arbitrarily high reliability without reducing the transmission rate to zero. The "fundamental theorem of information theory" states that this objective can be achieved by an optimal encoding strategy (the one that we considered here is very crude, since it amounts to simply re-sending each message $2n + 1$ times). It can be shown that the transmission rate in this case needs to be reduced to a value known as "channel capacity", but no lower than that.[2]

1.1.1 The Huffman Code

To get a sense for what an "efficient" encoding might look like, consider the fact that the average word in the English language consists of 4.5 letters. If we include spaces as well (which is necessary in order to separate words), this number increases to 5.5 characters. In light of this observation, what would be a suitable way to encode an English text? Since the alphabet consists of 27 letters (plus 1 space), the most obvious strategy would be to represent each character using 5 bits. Note, however, that this approach entails a certain redundancy, since four of the $2^5 = 32$ possible combinations will be unutilized. If we were to adopt such an encoding, the average number of bits per word would be $5.5 \times 5 = 27.5$.

Alternatively, we could focus our attention on the $16,356$ most "useful" words in the English language. If we were to do so, we could represent each of these words (as well as 27 letters and a space) using $2^{14} = 16,384$ different symbols. In this case, the average number of bits per word is reduced from 27.5 to only 14, but there is a possibility that we might encounter a word that we will not be able to encode.

The first approach (which allows us to spell out the message one letter at a time) is clearly more flexible, and covers *all* possibilities. If, on the other hand, we decide to sacrifice generality and focus only on the *most probable* messages, we can dramatically reduce their average length (measured in bits). The efficiency of the latter strategy is due to the fact that not all combinations of letters are equally likely in a language.

If we know the probabilities of different messages, one of the most effective ways to encode them is the so called *Huffman algorithm*. The following simple example illustrates how this scheme works.

Example 1.1. Suppose that the source can produce 8 different messages, $\{x_1, x_2, \ldots, x_8\}$, whose probabilities are given in Table 1.3.

Message	Probability
x_1	0.50
x_2	0.15
x_3	0.12
x_4	0.10
x_5	0.04
x_6	0.04
x_7	0.03
x_8	0.02

Table 1.3. Probabilities of messages $x_1 - x_8$.

In order to identify an optimal encoding strategy, we first need to arrange the messages in order of their likelihood. We then pick the two with the *lowest* probability, and connect them in the manner shown in Fig. 1.2. It is important to recognize that the "weight" of the new node represents the *sum* of these two probabilities, and that the branch with the *higher probability* is always assigned a 1 (in this diagram, such a branch is represented by a solid line).

The procedure described above is repeated recursively until we obtain a connected graph. Note that in every step we consider only the probabilities of the "active" nodes (i.e., nodes that have no outgoing edges), and assign a 1 and a 0 to the new pair of edges. The situation that arises after three steps of the process is shown in Fig. 1.3. At this point, four of the original nodes (x_1, x_2, x_3 and x_4), still have no outgoing edges and retain their original probabilities. There is also a new active node, whose probability is 0.13. It is hopefully clear that the next step involves connecting x_3 and x_4, since these two have the lowest probabilities among all remaining active nodes.

1.1. ENCODING AND ERROR PROBABILITIES

Figure 1.2: The first step of the Huffman algorithm.

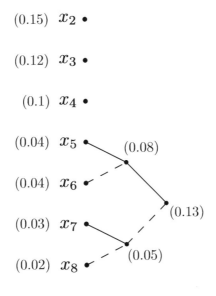

Figure 1.3: The situation after Step 3.

The situation after the final step is shown in Fig. 1.4. In order to associate an appropriate binary string with each message, we now "backtrack" from the point where $p = 1$ to the message that we are interested in. Such an assignment is unique, since there is only one possible path to any of the original nodes x_1, $x_2, \ldots, x_8$. The diagram in Fig. 1.5 indicates how the corresponding edge weights determine the encoding for message x_5 (which is 01011).

The overall scheme is shown in Table 1.4, where p_i denotes the probability of each message, and N_i represents the number of bits used to encode it. It is easily verified that the average number of bits per message is

$$\bar{N} = \sum_{i=1}^{8} N_i p_i = 2.26 \tag{1.17}$$

This is clearly better than an encoding which utilizes 3 bits for each x_i, which seems like a logical choice when the system produces 8 different messages. It turns out that the Huffman algorithm is actually the *most efficient* possible encoding when the probabilities p_i are different (this property can be shown rigorously).

Message	p_i	Encoding	N_i
x_1	0.50	1	1
x_2	0.15	011	3
x_3	0.12	001	3
x_4	0.10	000	3
x_5	0.04	01011	5
x_6	0.04	01010	5
x_7	0.03	01001	5
x_8	0.02	01000	5

Table 1.4. The Huffman encoding for messages $\{x_1, \ldots, x_8\}$.

1.2 Information Entropy

The notion of *uncertainty* plays a fundamental role in information theory, and provides a way to measure the amount of information that is transferred. In principle, it would be correct to say that we acquire more information about the source if the set of possible messages that it can produce is large. Under such circumstances, there is a significant *reduction* in uncertainty once a particular message is received.

1.2. INFORMATION ENTROPY

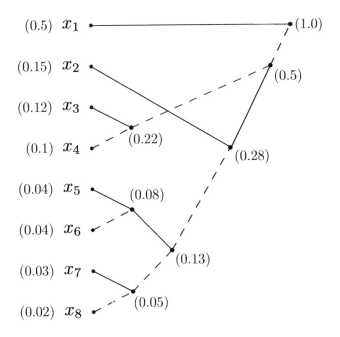

Figure 1.4: The final configuration.

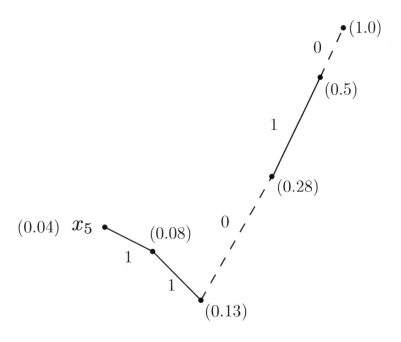

Figure 1.5: Huffman encoding for message x_5.

In order to quantify the degree of uncertainty that is associated with a given source, it makes sense to introduce some sort of "uncertainty measure" which satisfies a set of appropriate requirements. Defining such a function poses two separate (but related) challenges, the first of which is to identify a set of desirable mathematical properties that can adequately characterize the transfer of information. Once this is accomplished, we must consider whether there are functions that meet all of these requirements.

One way to approach this problem is to introduce a set of axioms that an "uncertainty function" must satisfy. Each of these axioms should make intuitive sense, and reflect the notion that information is proportional to the amount of uncertainty that is removed once a particular message is received. In order to formally describe this process, let us consider a random variable X, which can take values $\{x_1, x_2, \ldots, x_M\}$ with probabilities $\{p_1, p_2, \ldots, p_M\}$. With each outcome x_i we will associate a function $h(p_i)$, which represents *the amount of uncertainty that has been removed* once we establish that $X = x_i$ (in the following, we will refer to it as the "uncertainty associated with outcome x_i"). It is important to keep in mind that this function is the *same* for all types of experiments, and depends only on the corresponding probability p_i. This allows us to apply it to a very broad range of random processes.

In addition to function $h(p_i)$, we will also define the function

$$H_M(p_1, p_2, \ldots, p_M) = \sum_{i=1}^{M} p_i h(p_i) \qquad (1.18)$$

which represents the *average* (or *expected*) *uncertainty* associated with the entire set of possible outcomes $\{x_1, x_2, \ldots, x_M\}$. When working with this quantity, it is common practice to drop the subscript M (which indicates the number of possible outcomes), and use the symbol $H(X)$ to denote the "uncertainty of X", or the *information entropy*. In the following, we will use these two terms interchangeably.

At this point, we will not get into a detailed discussion of the axioms that $H(X)$ needs to satisfy. For our purposes, it suffices to say that the only function that meets all of these requirements has the form

$$H(p_1, p_2, \ldots, p_M) = -C \sum_{i=1}^{M} p_i \log(p_i) \qquad (1.19)$$

where C is an arbitrary constant, and the logarithm base is any number greater than one. The standard approach in communication theory is to set $C = 1$ and use base 2 logarithms, in which case the appropriate units for H are bits.

The function $H(X)$ defined in (1.18) has a number of important properties, which we will now examine. Before we do that, however, it is helpful to prove

1.2. INFORMATION ENTROPY

the following Lemma, which we will use to establish an upper bound on $H(X)$.

Lemma 1.1. Let $\{p_1, p_2, \ldots, p_M\}$ and $\{q_1, q_2, \ldots, q_M\}$ be two arbitrary sets of positive numbers, such that

$$\sum_{i=1}^{M} p_i = \sum_{i=1}^{M} q_i = 1 \tag{1.20}$$

Then,

$$-\sum_{i=1}^{M} p_i \log p_i \leq -\sum_{i=1}^{M} p_i \log q_i \tag{1.21}$$

Proof. We begin by observing that the function $f(x) = \log x$ satisfies

$$\log x \leq x - 1 \tag{1.22}$$

for *any* choice of x. Since this must hold for $x = q_i/p_i$ as well, we have that

$$\log\left(\frac{q_i}{p_i}\right) \leq \frac{q_i}{p_i} - 1 \tag{1.23}$$

Multiplying both sides by p_i and summing up, we obtain

$$\sum_{i=1}^{M} p_i \log\left(\frac{q_i}{p_i}\right) \leq \sum_{i=1}^{M}(q_i - p_i) = \sum_{i=1}^{M} q_i - \sum_{i=1}^{M} p_i = 0 \tag{1.24}$$

This implies that

$$\sum_{i=1}^{M} p_i(\log q_i - \log p_i) = \sum_{i=1}^{M} p_i \log q_i - \sum_{i=1}^{M} p_i \log p_i \leq 0 \tag{1.25}$$

which is what we set out to prove. **Q.E.D.**

We can now utilize Lemma 1.1 to prove the following simple result.

Theorem 1.3. The entropy function $H(p_1, p_2, \ldots, p_M)$ satisfies

$$H(p_1, p_2, \ldots, p_M) \leq \log M \tag{1.26}$$

Proof. Let us first observe that setting $q_i = 1/M$ ($i = 1, 2, \ldots, M$) in (1.21) yields

$$-\sum_{i=1}^{M} p_i \log p_i \leq -\sum_{i=1}^{M} p_i \log\left(\frac{1}{M}\right) = -\log\left(\frac{1}{M}\right)\sum_{i=1}^{M} p_i = \log M \tag{1.27}$$

since the elements of set $\{q_1, q_2, \ldots, q_M\}$ satisfy

$$\sum_{i=1}^{M} q_i = 1 \qquad (1.28)$$

Recalling that $H(p_1, p_2, \ldots, p_M)$ is defined as

$$H(p_1, p_2, \ldots, p_M) = -\sum_{i=1}^{M} p_i \log p_i \qquad (1.29)$$

inequality (1.26) follows directly. **Q.E.D.**

1.2.1 Joint Entropy

When the system under consideration contains multiple sources or allows for different types of observations, it makes sense to characterize it using several random variables. For the sake of simplicity, in the following we will assume that there are only two such variables, X and Y, which can take on values $\{x_1, x_2, \ldots, x_M\}$ and $\{y_1, y_2, \ldots, y_L\}$, respectively. With each possible outcome, we can then associate a *joint probability*

$$p(x_i, y_j) \equiv \Pr\left(X = x_i \text{ and } Y = y_j\right) \qquad (1.30)$$

and define the corresponding *joint entropy function* as

$$H(X, Y) = -\sum_{i=1}^{M}\sum_{j=1}^{L} p(x_i, y_j) \log p(x_i, y_j) \qquad (1.31)$$

It is not difficult to see that $H(X,Y) = H(Y,X)$ by definition, since $p(x_i, y_j) = p(y_j, x_i)$.

The following theorem establishes a simple connection between $H(X,Y)$ and the individual entropies $H(X)$ and $H(Y)$.

Theorem 1.4. The joint entropy satisfies

$$H(X, Y) \leq H(X) + H(Y) \qquad (1.32)$$

and (1.32) becomes an equality if and only if variables X and Y are *independent*.

Proof. We begin by observing that $p(x_i)$ can be viewed as the joint probability that $X = x_i$ and that Y takes on *any* possible value. We can formally express this as

$$p(x_i) = \sum_{j=1}^{L} p(x_i, y_j) \qquad (1.33)$$

1.2. INFORMATION ENTROPY

which holds true regardless of whether or not X and Y are independent. Similarly, we have that

$$p(y_j) = \sum_{i=1}^{M} p(x_i, y_j) \tag{1.34}$$

Expressions (1.33) and (1.34) allow us rewrite entropies $H(X)$ and $H(Y)$ in terms of joint probabilities, as

$$H(X) = -\sum_{i=1}^{M} p(x_i) \log p(x_i) = -\sum_{i=1}^{M} \left[\sum_{j=1}^{L} p(x_i, y_j) \right] \log p(x_i) \tag{1.35}$$

and

$$H(Y) = -\sum_{j=1}^{L} p(y_j) \log p(y_j) = -\sum_{j=1}^{L} \left[\sum_{i=1}^{M} p(x_i, y_j) \right] \log p(y_j) \tag{1.36}$$

Adding (1.35) and (1.36), we obtain

$$H(X) + H(Y) = -\sum_{i=1}^{M} \sum_{j=1}^{L} p(x_i, y_j) [\log p(x_i) + \log p(y_i)] =$$
$$= \sum_{i=1}^{M} \sum_{j=1}^{L} p(x_i, y_j) \log [p(x_i) p(y_j)] \tag{1.37}$$

Setting $p_{ij} \equiv p(x_i, y_j)$ and $q_{ij} \equiv p(x_i) p(y_j)$, we can rewrite $H(X, Y)$ and $H(X) + H(Y)$ in a simplified form, as

$$H(X, Y) = -\sum_{i=1}^{M} \sum_{j=1}^{L} p_{ij} \log p_{ij} \tag{1.38}$$

and

$$H(X) + H(Y) = -\sum_{i=1}^{M} \sum_{j=1}^{L} p_{ij} \log q_{ij} \tag{1.39}$$

Since

$$\sum_{i=1}^{M} \sum_{j=1}^{L} q_{ij} = \sum_{i=1}^{M} p(x_i) \sum_{j=1}^{L} p(y_j) = 1 \tag{1.40}$$

and

$$\sum_{i=1}^{M} \sum_{j=1}^{L} p_{ij} = \sum_{i=1}^{M} \sum_{j=1}^{L} p(x_i, y_j) = 1 \tag{1.41}$$

satisfy condition (1.20), we directly obtain

$$H(X,Y) = -\sum_{i=1}^{M}\sum_{j=1}^{L} p_{ij} \log p_{ij} \leq -\sum_{i=1}^{M}\sum_{j=1}^{L} p_{ij} \log q_{ij} = H(X) + H(Y) \quad (1.42)$$

by virtue of Lemma 1.1. Note also that $p_{ij} = p(x_i, y_j) = p(x_i)p(y_j) = q_{ij}$ when X and Y are independent variables. In that case, expression (1.42) obviously becomes an *equality*. **Q.E.D.**

1.2.2 Conditional Entropy

In situations when we have partial information about the system, it is useful to introduce the notion of *conditional entropy*. To see what that means, suppose once again that our system is characterized by two random variables, X and Y, and that we happen to know that $X = x_i$. Under such circumstances, the conditional entropy associated with Y can be defined as

$$H(Y|X = x_i) = -\sum_{j=1}^{L} p(y_j|x_i) \log p(y_j|x_i) \quad (1.43)$$

The *expected uncertainty* of Y (given that X will be revealed) can then be computed as

$$H(Y|X) = \sum_{i=1}^{M} p(x_i) H(Y|X = x_i) \quad (1.44)$$

Substituting (1.43) into (1.44), we obtain

$$H(Y|X) = -\sum_{i=1}^{M} p(x_i) \sum_{j=1}^{L} p(y_j|x_i) \log p(y_j|x_i) \quad (1.45)$$

We should recall at this point that $\Pr(X = x_i \text{ and } Y = y_j)$ is equal to the probability that we will first record $X = x_i$, and that we will subsequently observe $Y = y_j$ (given that $X = x_i$ is already known). This relationship is well known in probability theory, and can be formally expressed as

$$p(x_i, y_j) = p(y_j|x_i) p(x_i) \quad (1.46)$$

In light of expression (1.46), (1.45) can be rewritten as

$$H(Y|X) = -\sum_{i=1}^{M}\sum_{j=1}^{L} p(x_i, y_j) \log p(y_j|x_i) \quad (1.47)$$

1.2. INFORMATION ENTROPY

The following theorem shows how $H(Y|X)$ and $H(X|Y)$ are related.

Theorem 1.5. The joint and conditional entropies are related as

$$H(X,Y) = H(X) + H(Y|X) = H(Y) + H(X|Y) \qquad (1.48)$$

Proof. We begin by recalling that

$$H(X,Y) = -\sum_{i=1}^{M}\sum_{j=1}^{L} p(x_i, y_j) \log p(x_i, y_j) \qquad (1.49)$$

Using expression (1.46), we can rewrite the logarithm as

$$\log p(x_i, y_j) = \log[p(y_j|x_i)p(x_i)] = \log p(x_i) + \log p(y_j|x_i) \qquad (1.50)$$

which decomposes $H(X,Y)$ into two terms:

$$H(X,Y) = -\sum_{i=1}^{M} \log p(x_i) \sum_{j=1}^{L} p(x_i, y_j) - \sum_{i=1}^{M}\sum_{j=1}^{L} p(x_i, y_j) \log p(y_j|x_i) \qquad (1.51)$$

Observing that

$$\sum_{j=1}^{L} p(x_i, y_j) = p(x_i) \qquad (1.52)$$

this becomes

$$H(X,Y) = -\sum_{i=1}^{M} p(x_i) \log p(x_i) - \sum_{i=1}^{M}\sum_{j=1}^{L} p(x_i, y_j) \log p(y_j|x_i) = \qquad (1.53)$$

$$= H(X) + H(Y|X)$$

which is what we set out to prove. **Q.E.D.**

We now proceed to prove another useful result, which tells us that knowledge of X can decrease the uncertainty associated with Y only if the two variables are *correlated* in some way.

Theorem 1.6. Conditional entropy satisfies the inequality

$$H(Y|X) \leq H(Y) \qquad (1.54)$$

Expression (1.54) becomes an equality if and only if X and Y are independent.

Proof. According to Theorem 1.5, we have that

$$H(X) + H(Y|X) = H(X,Y) \tag{1.55}$$

On the other hand, Theorem 1.4 implies that

$$H(X,Y) \leq H(X) + H(Y) \tag{1.56}$$

Combining (1.55) and (1.56), we have that

$$H(X) + H(Y|X) \leq H(X) + H(Y) \tag{1.57}$$

which is equivalent to

$$H(Y|X) \leq H(Y) \tag{1.58}$$

In the special case when X and Y are *independent* variables, Theorem 1.4 implies that

$$H(X,Y) = H(X) + H(Y) \tag{1.59}$$

Substituting this into (1.55), we obtain

$$H(X) + H(Y|X) = H(X) + H(Y) \tag{1.60}$$

and therefore

$$H(Y|X) = H(Y) \tag{1.61}$$

Under such circumstances, the uncertainty associated with Y will remain unaffected by our knowledge of X. **Q.E.D.**

In systems that are characterized by multiple random variables, it is often of interest to compute the difference

$$I(X|Y) = H(X) - H(X|Y) \tag{1.62}$$

To get a sense for what this quantity means, we should recall that $H(X|Y)$ represents the uncertainty about X after Y is disclosed. Since $H(X)$ is the uncertainty associated with X before we knew anything about Y, it makes sense to interpret $I(X|Y)$ as the *information conveyed about X by Y* (this quantity is often referred to as *mutual information*). Note that $I(X|Y) = 0$ when X and Y are independent variables (by virtue of (1.61)).

The following simple lemma establishes an identity that allows us to compute $I(X|Y)$ if we are given probabilities $p(x_i)$, $p(y_i)$ and $p(x_i, y_j)$ (for $i, j = 1, 2, \ldots, M$).

Lemma 1.2. *The information conveyed about X by Y can be expressed as*

$$I(X|Y) = \sum_{i=1}^{M} \sum_{j=1}^{L} p(x_i, y_j) \log \frac{p(x_i, y_j)}{p(x_i) p(y_j)} \tag{1.63}$$

1.2. INFORMATION ENTROPY

Proof. We begin by observing that

$$H(X,Y) = H(X|Y) + H(Y) \tag{1.64}$$

according to Theorem 1.5. Substituting this into (1.62) we have

$$I(X|Y) = H(X) + H(Y) - H(X,Y) \tag{1.65}$$

Using expressions (1.29) and (1.31), (1.65) can be rewritten as

$$I(X|Y) = -\sum_{i=1}^{M} p(x_i) \log p(x_i) - \sum_{j=1}^{L} p(y_j) \log p(y_j) + \\ + \sum_{i=1}^{M}\sum_{j=1}^{L} p(x_i,y_j) \log p(x_i,y_j) \tag{1.66}$$

If we now recall that

$$p(x_i) = \sum_{j=1}^{L} p(x_i, y_j) \tag{1.67}$$

and

$$p(y_j) = \sum_{i=1}^{M} p(x_i, y_j) \tag{1.68}$$

the first two terms in (1.66) become

$$-\sum_{i=1}^{M}\sum_{j=1}^{L} p(x_i, y_j) \log p(x_i) \tag{1.69}$$

and

$$-\sum_{i=1}^{M}\sum_{j=1}^{L} p(x_i, y_j) \log p(y_j) \tag{1.70}$$

respectively. Combining these expressions, we now obtain

$$I(X|Y) = \sum_{i=1}^{M}\sum_{j=1}^{L} p(x_i, y_j) \log \frac{p(x_i, y_j)}{p(x_i)p(y_j)} \tag{1.71}$$

which is what we set out to prove. **Q.E.D.**

1.3 Interpretations of Information Entropy

The notion of information entropy has several possible interpretations, each of which provides us with a different insight into the nature of function $H(X)$. The simplest one is related to the quantity

$$W(p_i) = \log\left(\frac{1}{p_i}\right) \qquad (1.72)$$

which represents the *information content* of a message x_i whose probability is p_i. To see why information content is defined in this way, it suffices to observe that we learn nothing new if $p_i = 1$, since we know for sure that we will receive x_i. As a result, it makes sense to assume that $W(1) = 0$. On the other hand, one would expect to gain a great deal of new information when p_i approaches zero, since receiving an unlikely message constitutes a *surprise*. It is therefore reasonable to require that $W(p_i) \to \infty$ when $p_i \to 0$.

The connection between $W(p_i)$ and the entropy function H becomes apparent if we observe that

$$H = -\sum_{i=1}^{M} p_i \log(p_i) = \sum_{i=1}^{M} p_i \log W(p_i) \qquad (1.73)$$

can be understood as the *expected information content* of the message. When seen from that perspective, information entropy represents a measure for how "revealing" the received message is likely to be - the higher its value, the more we stand to learn from discovering which x_i has been transmitted.

1.3.1 Entropy and Encoding

An alternative interpretation of information entropy is related to the problem of encoding messages. In this context, H can be thought of as *the average number of Yes/No questions that need to be asked in order to identify which message has been sent*. To see the logic behind this interpretation, let us consider the case when the source can transmit five possible values, $\{x_1, x_2, \ldots, x_5\}$, with probabilities $p_1 = 0.3$, $p_2 = 0.2$, $p_3 = 0.2$, $p_4 = 0.15$ and $p_5 = 0.15$, respectively. The Huffman encoding of these messages is described in Fig. 1.6, and the schematic diagram in Fig. 1.7 shows how this scheme can be translated into a series of Yes/No questions. These results are summarized in Table 1.5, which tells us that the number of bits needed to encode a particular message x_i is *identical* to the number of Y/N questions that are needed to discover its identity.

1.3. INTERPRETATIONS OF INFORMATION ENTROPY   21

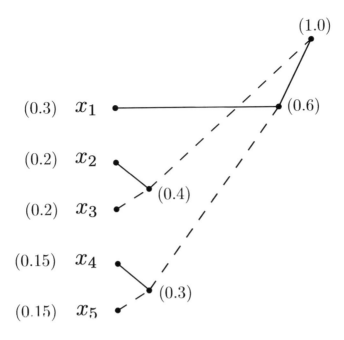

Figure 1.6: Huffman encoding for messages $x_1 - x_5$.

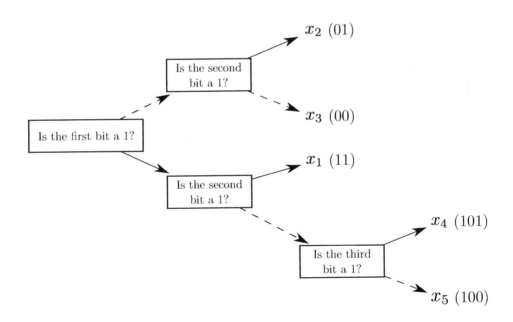

Figure 1.7: Y/N questions needed to identify each message.

It is not difficult to see that in this case the average number of bits (and therefore Yes/No questions as well) is

$$Q = 2 \cdot (0.3 + 0.2 + 0.2) + 3 \cdot (0.15 + 0.15) = 2.3 \tag{1.74}$$

This number is only slightly higher than the entropy

$$H(p_1, p_2, \ldots, p_5) = -\sum_{i=1}^{5} p_i \log(p_i) = 2.27 \tag{1.75}$$

which suggests that these two quantities might be related.

Outcome	Encoding	Y/N Questions
x_1	11	2
x_2	01	2
x_3	00	2
x_4	101	3
x_5	100	3

Table 1.5. The encoding of messages $x_1 - x_5$.

In order to strengthen this claim, we need to examine whether it is possible to improve the encoding so that the average number of bits is reduced. It turns out that something like this can be accomplished if we wait until *several* messages are generated, and then broadcast them all at once. To see the advantages of such an approach, let us consider the case when the source produces only two possible messages, x_1 and x_2, whose probabilities are $p_1 = 0.7$ and $p_2 = 0.3$, respectively. If we choose to transmit the messages one at a time, each of them can be encoded using a single bit (x_1 as 0 and x_2 as 1). As a result, the average number of Y/N questions is

$$Q = 0.7 \cdot 1 + 0.3 \cdot 1 = 1 \tag{1.76}$$

This is obviously higher than the entropy, which is computed as

$$H = -0.7 \log(0.7) - 0.3 \log(0.3) = 0.8833 \tag{1.77}$$

Let us now consider the scenario where two messages are aggregated. In that case we have four possible outcomes, whose probabilities are shown in Table 1.6.

1.3. INTERPRETATIONS OF INFORMATION ENTROPY

Outcome	Probability
(x_1, x_1)	$p_1^2 = 0.49$
(x_1, x_2)	$p_1 p_2 = 0.21$
(x_2, x_1)	$p_2 p_1 = 0.21$
(x_2, x_2)	$p_2^2 = 0.09$

Table 1.6. Probabilities of aggregated messages.

If we use the Huffman scheme shown in Fig. 1.8, the different combinations can be encoded in the manner described in Table 1.7.

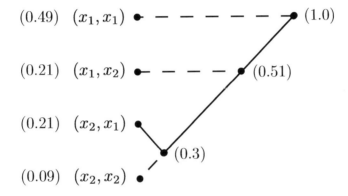

Figure 1.8: The Huffman algorithm for pairs of messages.

Outcome	Encoding
(x_1, x_1)	0
(x_1, x_2)	10
(x_2, x_1)	111
(x_2, x_2)	110

Table 1.7. Huffman encoding for the four possible pairs.

The number of Y/N questions needed to guess both outcomes *simultaneously* can be determined from the schematic diagram shown in Fig. 1.9. It is not difficult to see that this diagram matches Table 1.7 exactly. That leads us to conclude that the average number of bits needed to encode the *aggregated* message is

$$Q = 0.49 \cdot 1 + 0.21 \cdot (2 + 3) + 0.09 \cdot 3 = 1.81 \tag{1.78}$$

Remark 1.1. We could, of course, opt to encode the messages by using two bits for each pair (as shown in Table 1.8). This approach seems more straightforward than the Huffman algorithm, but it requires a higher average number of Y/N questions, since

$$Q = (0.49 + 0.21 + 0.21 + 0.09) \cdot 2 = 2 \tag{1.79}$$

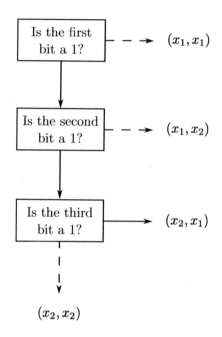

Figure 1.9: Y/N questions needed to identify each pair.

Outcome	Encoding
(x_1, x_1)	00
(x_1, x_2)	01
(x_2, x_1)	10
(x_2, x_2)	11

Table 1.8. An alternative encoding for the four pairs.

How should the value obtained in (1.78) be compared with the entropy computed in (1.77)? Since these two numbers aren't even remotely close, we need to find a way to account for the fact that Q corresponds to a *composite* message. The following theorem provides a result that will help us do so.

1.3. INTERPRETATIONS OF INFORMATION ENTROPY

Theorem 1.7. Suppose that a composite message consists of n individual messages x_i, each of which can take one of two possible values, 0 or 1. Assume further that the probabilities associated with these two values are p_1 and p_2, respectively. The entropy of the composite message can then be expressed as $H_T = nH$, where

$$H = -[p_1 \log p_1 + p_2 \log p_2] \tag{1.80}$$

represents the entropy associated with each individual message.

Proof. We begin by observing that an aggregated message consisting of k zeros and $n-k$ ones has probability

$$P(k) = p_1^k p_2^{n-k} \tag{1.81}$$

Given that there are

$$\binom{n}{k} = \frac{n!}{k!(n-k)!} \tag{1.82}$$

ways to form such a message, H_T can be expressed as

$$H_T = -\sum_{k=0}^{n} \binom{n}{k} P(k) \log P(k) = -\sum_{k=0}^{n} \binom{n}{k} p_1^k p_2^{n-k} \log(p_1^k p_2^{n-k}) \tag{1.83}$$

Recalling that

$$\log(p_1^k p_2^{n-k}) = k \log p_1 + (n-k) \log p_2 \tag{1.84}$$

the sum breaks up into two components,

$$H_1 = -\left[\sum_{k=0}^{n} k \binom{n}{k} p_1^k p_2^{n-k}\right] \cdot \log p_1 \tag{1.85}$$

and

$$H_2 = -\left[\sum_{k=0}^{n} (n-k) \binom{n}{k} p_1^k p_2^{n-k}\right] \cdot \log p_2 \tag{1.86}$$

In order to simplify the first sum, we will make use of the fact that

$$k \binom{n}{k} = \frac{k \cdot n!}{k!(n-k)!} = \frac{n!}{(k-1)!(n-k)!} \tag{1.87}$$

Since the term corresponding to $k = 0$ vanishes, the first sum becomes

$$H_1 = -\left[\sum_{k=1}^{n} \frac{n!}{(k-1)!(n-k)!} p_1^k p_2^{n-k}\right] \cdot \log p_1 \tag{1.88}$$

Setting $q = k - 1$, we now obtain

$$H_1 = -\left[\sum_{q=0}^{n-1} \frac{n!}{q!(n-q-1)!} p_1^{q+1} p_2^{n-q-1}\right] \cdot \log p_1 \qquad (1.89)$$

Factoring out n and p_1, (1.89) can be rewritten as

$$H_1 = -\left[\sum_{q=0}^{n-1} \frac{(n-1)!}{q!(n-q-1)!} p_1^q p_2^{(n-1)-q}\right] \cdot n p_1 \log p_1 \qquad (1.90)$$

It now remains to observe that

$$\sum_{q=0}^{n-1} \frac{(n-1)!}{q!(n-q-1)!} p_1^q p_2^{(n-1)-q} = \sum_{q=0}^{n-1} \binom{n-1}{q} p_1^q p_2^{(n-1)-q} = $$
$$= (p_1 + p_2)^{n-1} = 1 \qquad (1.91)$$

which implies that

$$H_1 = -n\, p_1 \log p_1 \qquad (1.92)$$

Following a similar procedure, we can simplify H_2 as well. We begin by observing that

$$(n-k)\binom{n}{k} = \frac{(n-k) \cdot n!}{k!(n-k)!} = \frac{n!}{k!(n-k-1)!} = n \cdot \binom{n-1}{k} \qquad (1.93)$$

Consequently,

$$H_2 = -\left[\sum_{k=0}^{n-1} \binom{n-1}{k} p_1^k p_2^{(n-1)-k}\right] \cdot n p_2 \log p_2 \qquad (1.94)$$

Since

$$\sum_{k=0}^{n-1} \binom{n-1}{k} p_1^k p_2^{(n-1)-k} = (p_1 + p_2)^{n-1} = 1 \qquad (1.95)$$

it follows that

$$H_2 = -n\, p_2 \log p_2 \qquad (1.96)$$

Combining (1.92) and (1.96), we can now relate H_T to the entropy of the original source (which produces only two types of messages - a zero or a one) as

$$H_T = H_1 + H_2 = -n\left[p_1 \log p_1 + p_2 \log p_2\right] = nH \qquad (1.97)$$

Q.E.D.

1.3. INTERPRETATIONS OF INFORMATION ENTROPY

Remark 1.2. This result remains unchanged if we allow x_i to take M possible values, with probabilities $\{p_1, p_2, \ldots, p_M\}$. In that case, we have

$$H_T = -n \sum_{k=1}^{M} p_k \log p_k = nH \tag{1.98}$$

Theorem 1.7 allows us to meaningfully compare the results obtained in (1.77) and (1.78), since the entropy of the *aggregated* message is

$$H_T = 2H = 2 \cdot 0.8833 = 1.7666 \tag{1.99}$$

It is readily observed that $Q = 1.81$ is a much better approximation of H_T than in the case of a single message (where we had $H = 0.8833$ and $Q = 1$). As a result, we can conclude that it is more efficient to send n messages together, and then try to guess all of the outcomes simultaneously.

Remark 1.3. In order to avoid dealing with H_T, it makes sense to compare H with the average number of bits *per message*, which is $\bar{Q} = Q/2 = 1.81/2 = 0.95$ in our example. Following this logic, we would obviously need to compare H with $\bar{Q} = Q/n$ in case the aggregate message consists of n individual ones. It can be shown that the number of bits per message, $\bar{Q}$, approaches $H(X)$ as $n \to \infty$ (which obviously justifies the idea of encoding *entire blocks* of messages instead of individual messages).

1.3.2 Entropy and Probable Sequences

In order to introduce the third interpretation of information entropy, let us once again consider the scenario in which n individual messages are aggregated into a single long message, which we will denote by α. In the following, we will assume that the source can produce M possible outcomes $\{x_1, x_2, \ldots, x_M\}$, with probabilities $\{p_1, p_2, \ldots, p_M\}$. We will also define functions $f_i(\alpha)$ ($i = 1, 2, \ldots, M$), which represent the number of times x_i appears in message α.

To illustrate what functions $f_i(\alpha)$ really mean, it is helpful to consider a simple example where $M = 5$ and the possible outcomes are $\{x_1, x_2, \ldots, x_5\}$. If we assume that the aggregated message consists of $n = 10$ individual messages, α might take the form

$$\alpha = \{x_2,\ x_5,\ x_1,\ x_5,\ x_2,\ x_3,\ x_2,\ x_1,\ x_4,\ x_5\} \tag{1.100}$$

in which case we have

$$f_1(\alpha) = 2$$
$$f_2(\alpha) = 3$$
$$f_3(\alpha) = 1 \qquad (1.101)$$
$$f_4(\alpha) = 1$$
$$f_5(\alpha) = 3$$

If the probabilities of the different outcomes are $\{p_1, p_2, \ldots, p_5\}$, we can express the probability of encountering this particular α as

$$\Pr(\alpha) = p_1^{f_1(\alpha)} p_2^{f_2(\alpha)} p_3^{f_3(\alpha)} p_4^{f_4(\alpha)} p_5^{f_5(\alpha)} = p_1^2 p_2^3 p_3 p_4 p_5^3 \qquad (1.102)$$

Note that the order in which the individual messages appear is not important in this case - all we are interested in at this time is the *frequency* with which different outcomes appear. From that perspective, it would be fair to say that α is, in fact, a "representative" for an entire class of aggregated messages.

It is not difficult to show that each function $f_i(\alpha)$ defined in this manner must have a binomial probability distribution. Indeed, the probability that $f_i(\alpha) = k$ is equal to the probability that we will encounter exactly k x_i's and $n - k$ other outcomes in α. Each such arrangement has probability

$$p(k) = p_i^k (1 - p_i)^{n-k} \qquad (1.103)$$

and there are exactly $\binom{n}{k}$ of them, which means that

$$\Pr\{f_i(\alpha) = k\} = \binom{n}{k} p_i^k (1 - p_i)^{n-k} \qquad (1.104)$$

Since the distribution is binomial, the expected value of $f_i(\alpha)$ and its variance can be computed as

$$\mu_i = n p_i \qquad (1.105)$$

and

$$\sigma_i^2 = n p_i (1 - p_i) \qquad (1.106)$$

These properties allow us to introduce the notion of a *typical sequence* α in the following manner.

Definition 1.1. For a given n, let ε_n be an arbitrary positive number, and let $k(\varepsilon_n)$ be an integer that satisfies

$$\frac{1}{k^2(\varepsilon_n)} < \frac{\varepsilon_n}{M} \qquad (1.107)$$

1.3. INTERPRETATIONS OF INFORMATION ENTROPY

We will say that α is a typical sequence if the inequality

$$\left| \frac{f_i(\alpha) - \mu_i}{\sigma_i} \right| = \left| \frac{f_i(\alpha) - np_i}{\sqrt{np_i(1-p_i)}} \right| < k(\varepsilon_n) \tag{1.108}$$

holds for $i = 1, 2, \ldots, M$.

Note that this definition depends on our choice of ε_n and $k(\varepsilon_n)$, and is therefore *non-unique*. Nevertheless, it will prove to be very useful in describing certain asymptotic properties of aggregated messages.

We now proceed to prove the following theorem, which establishes a connection between the entropy of the source and typical sequences.

Theorem 1.8. Let us assume that the source produces M possible outcomes $\{x_1, x_2, \ldots, x_M\}$, with probabilities $\{p_1, p_2, \ldots, p_M\}$. We will further assume that n individual messages are aggregated into a *single* message α of length n. The following three claims can then be made about such a message.

Property 1. The probability that α will be a *non-typical* sequence is lower than ε_n, where $\varepsilon_n \to 0$ when $n \to \infty$.

Property 2. If α is an arbitrary typical sequence, the probability that we will observe it can be bounded as

$$2^{-nH - A_n\sqrt{n}} < \Pr(\alpha) < 2^{-nH + A_n\sqrt{n}} \tag{1.109}$$

where

$$H = -\sum_{i=1}^{M} p_i \log p_i \tag{1.110}$$

is the entropy of the source, and A_n depends on $k(\varepsilon_n)$ and probabilities $\{p_1, p_2, \ldots, p_M\}$.

Property 3. The number of typical sequences of length n is $2^{n(H+\delta_n)}$, where $\lim_{n \to \infty} \delta_n = 0$.

Proof. In the following, we will prove each of the three properties separately.

Property 1. If α is a *non typical* sequence, it follows that inequality (1.108) must be violated for *at least* one i. The probability that this will occur can be expressed as

$$\sum_{i=1}^{M} \Pr\left\{ \left| \frac{f_i(\alpha) - np_i}{\sqrt{np_i(1-p_i)}} \right| \geq k(\varepsilon_n) \right\} \tag{1.111}$$

Recalling that $\mu_i = np_i$ and $\sigma_i = \sqrt{np_i(1-p_i)}$ represent the mean and the variance of random variable $f_i(\alpha)$, we can invoke Chebyshev's inequality,[3] which states that a random variable X with mean μ and variance σ satisfies

$$\Pr\{|X - \mu| \geq k\sigma\} \leq \frac{1}{k^2} \qquad (1.112)$$

for any $k > 0$. Replacing X and k with $f_i(\alpha)$ and $k(\varepsilon_n)$, respectively, we have

$$\Pr\left\{|f_i(\alpha) - np_i| \geq k(\varepsilon_n)\sqrt{np_i(1-p_i)}\right\} \leq \frac{1}{k^2(\varepsilon_n)} < \frac{\varepsilon_n}{M} \qquad (1.113)$$

(by virtue of (1.107)). Since this holds true for all i, it follows that

$$\sum_{i=1}^{M} \Pr\left\{|f_i(\alpha) - np_i| \geq k(\varepsilon_n)\sqrt{np_i(1-p_i)}\right\} < M \cdot \frac{\varepsilon_n}{M} = \varepsilon_n \qquad (1.114)$$

which is precisely what Property 1 claims.

It is important to note in this context that we can always choose ε_n in such a way that $\varepsilon_n \to 0$ while ensuring that $k(\varepsilon_n)$ grows more slowly than $\sqrt{n}$. As an illustration of this possibility, let us set $\varepsilon_n = n^{-0.3}$ and pick $k(\varepsilon_n)$ as $k(\varepsilon_n) = \sqrt{M}n^{0.2}$. In that case

$$\frac{1}{k^2(\varepsilon_n)} = \frac{1}{Mn^{0.4}} < \frac{1}{Mn^{0.3}} = \frac{\varepsilon_n}{M} \qquad (1.115)$$

and it is easily verified that

$$\frac{k(\varepsilon_n)}{\sqrt{n}} = \frac{\sqrt{M}n^{0.2}}{n^{0.5}} = \frac{\sqrt{M}}{n^{0.3}} \to 0 \quad \text{when} \quad n \to \infty \qquad (1.116)$$

Property 2. If α is a typical sequence, then

$$\left|\frac{f_i(\alpha) - np_i}{\sqrt{np_i(1-p_i)}}\right| < k(\varepsilon_n) \qquad (1.117)$$

holds by definition for $i = 1, 2, \ldots, M$. This inequality can be equivalently rewritten as

$$np_i - k(\varepsilon_n)\sqrt{np_i(1-p_i)} < f_i(\alpha) < np_i + k(\varepsilon_n)\sqrt{np_i(1-p_i)} \qquad (1.118)$$

We should recall at this point that the probability of encountering sequence α is given by an expression of the form (1.102). As a result, we have that

$$\log \Pr(\alpha) = \log\left(p_1^{f_1(\alpha)} \cdots p_M^{f_M(\alpha)}\right) = \sum_{i=1}^{M} f_i(\alpha)\log(p_i) \qquad (1.119)$$

1.3. INTERPRETATIONS OF INFORMATION ENTROPY

Using the bounds on $f_i(\alpha)$ established in (1.118), we further obtain

$$\sum_{i=1}^{M} \left[np_i - k(\varepsilon_n)\sqrt{np_i(1-p_i)} \right] \log(p_i) < \log \Pr(\alpha) \qquad (1.120)$$
$$< \sum_{i=1}^{M} \left[np_i + k(\varepsilon_n)\sqrt{np_i(1-p_i)} \right] \log(p_i)$$

Setting

$$A_n \equiv k(\varepsilon_n) \sum_{i=1}^{M} \sqrt{p_i(1-p_i)} \log(p_i) \qquad (1.121)$$

(1.120) becomes

$$n \sum_{i=1}^{M} p_i \log(p_i) - A_n \sqrt{n} < \log \Pr(\alpha) < n \sum_{i=1}^{M} p_i \log(p_i) + A_n \sqrt{n} \qquad (1.122)$$

Since

$$H = -\sum_{i=1}^{M} p_i \log(p_i) \qquad (1.123)$$

by definition, it follows that

$$-nH - A_n \sqrt{n} < \log \Pr(\alpha) < -nH + A_n \sqrt{n} \qquad (1.124)$$

and therefore

$$2^{-nH - A_n \sqrt{n}} < \Pr(\alpha) < 2^{-nH + A_n \sqrt{n}} \qquad (1.125)$$

To see what happens when $n \to \infty$, we can rewrite (1.125) as

$$2^{-n(H + A_n/\sqrt{n})} < \Pr(\alpha) < 2^{-n(H - A_n/\sqrt{n})} \qquad (1.126)$$

Observing that A_n has the form $A_n \equiv k(\varepsilon_n) \cdot C$ (where C is a constant) and recalling that $k(\varepsilon_n)$ can be chosen so that $k(\varepsilon_n)/\sqrt{n} \to 0$ when $n \to \infty$, it follows that $A_n/\sqrt{n} \to 0$ as well. As a result, we can conclude that

$$\Pr(\alpha) \to 2^{-nH} \qquad (1.127)$$

when $n \to \infty$.

Property 3. Let S denote the set of all typical sequences. By virtue of Property 1, the probability that α is a *non-typical* sequence is

$$\Pr(\alpha \notin S) = 1 - \Pr(\alpha \in S) < \varepsilon_n \qquad (1.128)$$

This allows us to bound $\Pr(\alpha \in S)$ as

$$1 - \varepsilon_n < \Pr(\alpha \in S) < 1 \qquad (1.129)$$

On the other hand, the probability that we will encounter a typical sequence is equal to the sum of the individual probabilities of every sequence that belongs to S

$$\Pr(\alpha \in S) = \sum_{i=1}^{N_S} P_i \qquad (1.130)$$

where N_S denotes the total number of elements in set S. We know from equation (1.125) that each P_i must satisfy

$$c \equiv 2^{-nH - A_n\sqrt{n}} < P_i < 2^{-nH + A_n\sqrt{n}} \equiv d \qquad (1.131)$$

(since it is a typical sequence). Substituting

$$\frac{P_i}{d} < 1 \qquad (1.132)$$

into (1.129) and (1.130), we obtain

$$\frac{1 - \varepsilon_n}{d} < \frac{\Pr(\alpha \in S)}{d} = \sum_{i=1}^{N_S} \frac{P_i}{d} < N_S \qquad (1.133)$$

If we now make use of the fact that

$$1 < \frac{P_i}{c} \qquad (1.134)$$

we have

$$N_S < \sum_{i=1}^{N_S} \frac{P_i}{c} = \frac{1}{c} \sum_{i=1}^{N_S} P_i \leq \frac{1}{c} \qquad (1.135)$$

Combining inequalities (1.133) and (1.135) produces

$$(1 - \varepsilon_n) 2^{nH - A_n\sqrt{n}} = \frac{1 - \varepsilon_n}{d} < N_S < \frac{1}{c} = 2^{nH + A_n\sqrt{n}} \qquad (1.136)$$

Observing that

$$(1 - \varepsilon_n) 2^{nH - A_n\sqrt{n}} = 2^{nH - A_n\sqrt{n} + \log(1-\varepsilon_n)} =$$
$$= 2^{n(H - A_n n^{-1/2} + n^{-1} \log(1-\varepsilon_n))} \qquad (1.137)$$

and setting $\delta_n^{(1)} = -A_n/\sqrt{n} + n^{-1} \log(1 - \varepsilon_n)$, we obtain

$$2^{n(H + \delta_n^{(1)})} < N_S \qquad (1.138)$$

1.3. INTERPRETATIONS OF INFORMATION ENTROPY

We also know that

$$2^{nH+A_n\sqrt{n}} = 2^{n(H+A_n/\sqrt{n})} = 2^{n(H+\delta_n^{(2)})} \qquad (1.139)$$

where $\delta_n^{(2)} = A_n/\sqrt{n}$. Combining (1.138) and (1.139), we now obtain

$$2^{n(H+\delta_n^{(1)})} < N_S < 2^{n(H+\delta_n^{(2)})} \qquad (1.140)$$

Since $A_n/\sqrt{n} \to 0$ when $n \to \infty$, it follows that both $\delta_n^{(1)}$ and $\delta_n^{(2)}$ become vanishingly small when n is sufficiently large. As a result, we can conclude that $N_S \to 2^{nH}$ when $n \to \infty$. **Q.E.D.**

Theorem 1.8 allows us to draw several important conclusions about the asymptotic behavior of aggregated messages. Perhaps the most important one is that Properties 2 and 3 hold for practically *all* long sequences, since the likelihood of encountering a non-typical one approaches zero when $n \to \infty$ (according to Property 1). This means that the number of such sequences can be estimated as 2^{nH}, and that each of them occurs with probability 2^{-nH}. Equation (1.118) also indicates that

$$\lim_{n \to \infty} \frac{f_i(\alpha)}{n} = p_i \qquad (1.141)$$

since $k(\varepsilon_n)/\sqrt{n} \to 0$ when $n \to \infty$. This implies that each x_i appears approximately np_i times in a long aggregated message.

The practical significance of this result lies in the fact that only 2^{nH} out of M^n possible sequences are *likely* (when n is sufficiently large, of course). All other sequences have vanishingly small probabilities of being observed, and can therefore be safely ignored. Entropy provides us with a way to explicitly describe this distinction, and leads us to a third possible interpretation of H - we can interpret it *as a measure of the probability that a given message of length n will be recorded* (as shown in Theorem 1.8, this probability approaches 2^{-nH} as $n \to \infty$, and is the same for all messages).

Chapter 2

Information in Biology

2.1 Protein Synthesis

Biological information "resides" in the DNA, which is a nucleic acid whose basic "building blocks" are four relatively simple molecular sub-units known as *nucleotides*. These sub-units are composed of an organic base, a sugar molecule (deoxyribose) and a phosphoric acid, and are usually described by the letters A, G, C and T (which stand for adenine, guanine, cytosine and thymine). RNA represents a closely related class of nucleic acids, whose chemical composition is similar to that of DNA. The two differ in the type of sugar molecule (RNA contains ribose instead of deoxyribose), and the fact that in RNA thymine (T) is replaced by uracil (U).

Nucleotides G and C have a chemical affinity to bond, as do A and T (or A and U in the case of RNA). This property allows DNA molecules to *self-replicate* in a fairly straightforward manner. In the first step of this process, each nucleotide bonds with its *complementary* nucleotide (G with C, and A with T), which produces a "negative" image of the original. The "negative" then splits off, and provides the template for a "positive" copy. This copy needn't always be exact, however, since the affinity between C and G (or A and T) does not rule out the possibility of occasional "incorrect" pairings. This sort of "copying error" is one of the ways in which genetic mutations arise.

The sequence of nucleotides in DNA molecules contains all the information needed for the synthesis of proteins, which are essential for the proper functioning of a living cell. Proteins are made up of 20 different types of *amino acids*, which are arranged in a linear sequence. These 20 amino acids constitute the "alphabet" of protein language, which is considerably more complex than the basic four-character "alphabet" of DNA. The analysis of protein molecules is further complicated by the fact that they have a folding three dimensional structure, which allows for interactions between segments that are far apart in the linear arrangement.

The chemical structure of a protein is determined by triplets of nucleotides, which are known as *codons*. Each codon corresponds to a particular amino acid, but the mapping is not "one-to-one". This is due to the fact that there are only 20 different types of amino acids, and the four types of nucleotides (A, T, G and C) can be combined to produce as many as $4^3 = 64$ different triplets.

The correspondence between codons and amino acids is shown in Table 2.1. Note that most amino acids correspond to multiple codons, which indicates that there is a certain amount of redundancy in the way triplets of nucleotides are mapped. Only Methionine and Tryptophan have a unique encoding, and can be precisely associated with a particular codon (AUG and UGG, respectively).

Amino Acid	Abbreviation	Triplets of Nucleotides
Glycine	gly	{GGG, GGC, GGU, GGA}
Proline	pro	{CCG, CCC, CCU, CCA}
Leucine	leu	{CUG, CUC, CUU, CUA, UUA, UUG}
Arginine	arg	{CGG, CGC, CGU, CGA}
Threonine	thr	{ACG, ACC, ACU, ACA}
Valine	val	{GUG, GUC, GUU, GUA}
Alanine	ala	{GCG, GCC, GCU, GCA}
Serine	ser	{UCG, UCC, UCU, UCA}
Phenylalanine	phe	{UUU, UUC}
Cysteine	cys	{UGU, UGC}
Glutamine	gln	{CAA, CAG}
Aspartic acid	asp	{AAU, AAC}
Glutamic acid	glu	{GAA, GAG}
Isoleucine	ile	{AUU, AUC, AUA}
Tryptophan	trp	{UGG}
Histidine	his	{CAU, CAC}
Lysine	lys	{AAA, AAG}
Asparagine	asn	{GAU, GAC}
Methionine	met	{AUG}
Tyrosine	tyr	{UAU, UAC}
Unmatched		{UAA, UAG, UGA}

Table 2.1. Correspondence between codons and amino acids.

Redundancies in encoding inevitably result in a loss of information, since we cannot recover the exact sequence of nucleotides by observing the structure of the protein that they produced. This needn't be viewed as a deficiency, however, since redundancy also helps reduce the impact of "copying errors".

2.1. PROTEIN SYNTHESIS

To see why this is so, consider the mapping in Table 2.2, which shows how the US zip code is encoded in binary form.

In this mapping, the source "alphabet" consists of $2^5 = 32$ different strings of ones and zeros, while the "alphabet" on the receiving end has only 10. As a result, only 10 of the possible messages are meaningful, while the remaining 22 represent "nonsense". Note, however, that this apparent redundancy allows us to choose the binary representation of numbers $0-9$ in such a way that each one differs from the others in *at least two positions*. As a result, an error in a single bit will always produce a "nonsensical" combination, which cannot be confused with another digit. In such a system, an incorrect interpretation would require *two* errors, which is highly improbable.

$$
\begin{aligned}
0 &\longleftrightarrow 11000 \\
1 &\longleftrightarrow 00011 \\
2 &\longleftrightarrow 00101 \\
3 &\longleftrightarrow 00110 \\
4 &\longleftrightarrow 01001 \\
5 &\longleftrightarrow 01010 \\
6 &\longleftrightarrow 01100 \\
7 &\longleftrightarrow 10001 \\
8 &\longleftrightarrow 10010 \\
9 &\longleftrightarrow 10100
\end{aligned}
$$

Table 2.2. An encoding that minimizes the likelihood of errors.

Unlike the ZIP code (where roughly two thirds of the possible messages fall into the "nonsense" category), almost all of the 64 possible triplets of nucleotides represent meaningful combinations, the only exceptions being UAA, UAG and UGA. These three "unmatched" codons are not completely useless, however, since their appearance in the sequence signals that the translation of RNA into proteins should stop.

The synthesis of proteins takes place in a part of the cell known as the *ribosome*. To get a sense for how this process works, consider the following hypothetical DNA sequence, which has been partitioned into triplets of nucleotides for convenience.

$$\cdots \quad G\ T\ C \ \leftrightarrow\ G\ C\ A \ \leftrightarrow\ A\ G\ C \ \leftrightarrow\ G\ T\ T \quad \cdots$$

This sequence is first transcribed into *messenger* RNA (mRNA) by replacing all instances of T with U, which produces

$$\cdots \text{ G U C } \leftrightarrow \text{ G C A } \leftrightarrow \text{ A G C } \leftrightarrow \text{ G U U } \cdots$$

The triplets are then translated *from right to left*, following the scheme described in Table 2.1. The results of this transformation are shown in Table 2.3, from which we can conclude that the sequence of amino acids that corresponds to this segment of the DNA is: ... leu → arg → thr → leu ...[4]

Triplet		Amino acid
UUG	→	leu
CGA	→	arg
ACG	→	thr
CUG	→	leu

Table 2.3. "Translation" of the sample DNA segment.

2.2 Information Processing in Biological Systems

The transfer of biological information bears a certain resemblance to interpersonal communications, in the sense that here, too, we tend to be more interested in the meaning of the message than in its syntax. In the context of biology, the "meaning" of a message is usually understood in terms of the role that it plays in the development of a living organism. Unfortunately, we have not yet developed adequate mathematical models for describing such processes, and it is questionable whether this is even possible. It seems, therefore, that the mathematical tools currently available to us can only be used to analyze *syntactic* aspects of genetic information.

In order to adapt these tools to biological systems, we will first need to identify what constitutes the basic "building blocks" of a genetic message. There are two rather different possibilities in this case, which must be clearly distinguished. We could assume, for example, that a "message" represents a string of randomly arranged nucleotides, each of which has a particular probability of occurring. On the other hand, we could also think of genetic "messages" as sequences of amino acids, in which case the corresponding set of probabilities would be very different.

Once we decide which of these two possibilities is more appropriate for our purposes, we need to determine if (and how) the mathematical framework described in Chapter 1 can be applied to DNA sequences. As a first step in this process, we need to address the following two questions, which will help

2.2. INFORMATION PROCESSING IN BIOLOGICAL SYSTEMS

us establish whether nucleotides can be treated like random messages with different probabilities (as is commonly done in communication theory).

Question 1. Do the nucleotides that constitute the DNA sequence exhibit some sort of pattern, or should we treat them as a random collection of symbols?

Question 2. Is it reasonable to expect that all possible combinations of nucleotides are equally likely?

In dealing with the first question, it is helpful to consider what happens when we encounter a long sequence of symbols that shows no signs of regularity. It is certainly possible that such a sequence could, in fact, be random, but it could also be part of a periodic sequence which has a very long period. Telling the difference between the two can be a very difficult task, since proving the randomness of a sequence is equivalent to showing that the *shortest* program which produces it is as long as the sequence itself. This can be done in some very special cases, but not in general.[5]

What can we say about such sequences in the absence of a rigorous proof? One thing that we know for sure is that they are *very likely* to be random. To see why, let us examine all possible sequences of length n, and estimate how many of them are *not* random. In order to do that, we will assume that such a sequence is random if the length of the program that produces it is *at least* $n - 10$ bits (this "threshold" is, of course, somewhat arbitrary, since one could reasonably choose a different criterion). The number of possible programs whose length does not exceed $n - 11$ bits is not difficult to compute. Indeed, observing that there are 2^k possible programs whose length is k bits, we have that there are

$$N = \sum_{k=1}^{n-11} 2^k = 2^{n-10} - 2 \qquad (2.1)$$

programs whose length less than $n - 10$.

Even if we assume that *each* of these programs produces a different binary sequence of length n (which is by no means certain), the number of such sequences would only be a small fraction of the overall number of possibilities, since

$$\frac{2^{n-10} - 2}{2^n} \approx 2^{-10} \approx 10^{-3} \qquad (2.2)$$

when n is sufficiently large. We can conclude, in other words, that according to this criterion only one out of every thousand possible sequences is *not* random (and that would be the best case scenario).[6]

The second question that we posed asks us to evaluate whether all DNA sequences are *equally likely*. In order to do that, we should first note that the genomes of living organisms are very complex, and are made up of millions of nucleotides. The genome of the *E.coli* bacterium, for example, consists of some four million nucleotides, which gives rise to a staggering $4^{4,000,000}$ possible combinations. It turns out, however, that only a tiny fraction of these combinations actually occurs in nature, and carries information that is conducive to the proper functioning of a living cell.

To this we should add the observation that even those chains of nucleotides that are "biologically useful" contain a great deal of unnecessary information. Experimental evidence suggests that only about 10% of the DNA chain is actually used for regulatory purposes or for synthesizing proteins – the remaining 90% appears to have no function at all (which is why it is commonly referred to as "junk DNA").

Theorem 1.8 (which we proved in Section 1.3.2) provides us with an additional reason for assuming that certain sequences of nucleotides occur more frequently than others. This theorem tells us that when long strings of symbols are formed from M different characters whose probabilities are $\{p_1, p_2, \ldots p_M\}$, only a small subset of them is likely to be produced. This subset consists of 2^{nH} elements, where n denotes the number of symbols in the string, and

$$H = -\sum_{k=1}^{M} p_k \log_2 p_k \tag{2.3}$$

represents the information entropy.

To get a sense for what this means in practice, consider an experiment in which each element in the string of symbols is formed by throwing two dice. In this case, we will obviously have 11 possible symbols, since the two dice can produce values ranging from 2 to 12. Note, however, that these outcomes are *not* equally likely, since 2 and 12 can arise in only one way (and therefore have a probability of $p_2 = p_{12} = 1/36$), while other values correspond to multiple combinations. The number 7, for example, can arise in six different ways - as $1+6$, $2+5$, $3+4$, $4+3$, $5+2$ and $6+1$. As a result, its probability is

$$p_7 = 6/36 = 1/6 \tag{2.4}$$

Given that we know all the probabilities in this case, we can easily compute the entropy H, and therefore 2^{nH} as well for any given n. When n is sufficiently large, this number represents only a miniscule portion of the 11^n different sequences that can be formed in this way. We can see this very clearly if we set $n = 100$, in which case we have that

$$\frac{2^{nH}}{11^n} = 2.69 \cdot 10^{-6} \tag{2.5}$$

This means that less than three out of every million possible sequences are actually likely to occur.

2.3. REDUNDANCY AND "GENETIC NOISE"

Calculations of this sort have significant implications for molecular biology, since the 20 amino acids that make up proteins do not have the same probabilities of arising in nature. As a result, we have every reason to expect that only a minute fraction of the 20^n possible combinations will actually occur.

2.3 Redundancy and "Genetic Noise"

If we think of the genome as a collection of randomly produced symbols (A, C, T and G), then it is natural to compare the DNA to the *source* in a communication system. The process of transcription (in which the information contained in the DNA is translated into the language of mRNA) can then be viewed as a form of *encoding*, and genetic mutations can be interpreted as *noise* that corrupts the message as it passes through the communication channel. The genetic message is *decoded* in the ribosomes, at which point triplets of nucleotides are "translated" into amino acids. Once these amino acids are assembled into a protein, we can say that the message has been *received* and *interpreted*.

It is important to stress once again that in this process, the original message is converted from the mRNA "alphabet" (which consists of 64 "letters") into the protein "alphabet", whose "letters" are the 20 amino acids. This reduction in the size of the alphabet accounts for the *redundancy* that is introduced in the course of transmitting and decoding the DNA message. As noted earlier, redundancy can be quite useful in minimizing the effects of errors that arise due to "genetic noise". Indeed, if there was a 1-to-1 mapping between the mRNA and protein alphabets, any mutation would automatically change the resulting amino acid. However, since there are 44 "extra" triplets in the system, almost all amino acids have multiple possible sources, and the risk of encountering errors is significantly reduced.

A standard approach for minimizing errors in man-made communication systems has been to assign *greater* redundancy to those "letters" that occur more frequently. A number of authors have pointed out that the genetic code seems to be organized in a similar manner, since the most common amino acids do, in fact, have the largest number of corresponding triplets. It has been argued that this sort of optimization is the result of evolutionary pressures, which naturally produced a "code" that is resilient with respect to "genetic noise".

To see how the effects of redundancy and "genetic noise" can be described in mathematical terms, let X and Y denote a pair of random variables that correspond to *triplets of nucleotides* and *amino acids*, respectively. From a communications perspective, variable X (which has 64 possible values) can be viewed as the message that is being sent, and Y represents its interpretation on the receiving end (which can take one of 20 possible values).

Following equation (1.62), we can compute the *mutual entropy* of such a system as

$$I(X|Y) = H(X) - H(X|Y) \tag{2.6}$$

(as noted in Chapter 1, this quantity represents the information conveyed about X by Y). The term $H(X|Y)$ in (2.6) represents the *conditional entropy*, which reflects the remaining uncertainty about X once Y is known. In order to express this term in a suitable form, we should first observe that its "mirror image", $H(Y|X)$, can be computed as

$$H(Y|X) = -\sum_{i=1}^{M}\sum_{j=1}^{L} p(x_i)p(y_j|x_i) \log p(y_j|x_i) \tag{2.7}$$

(see equation (1.45) in Section 1.2.2). In this expression, variables $\{x_i\}$ denote individual triplets of nucleotides, and $\{y_j\}$ correspond to amino acids. The probabilities that are associated with these variables are $p(x_i)$ and $p(y_j)$, respectively, and $p(y_j|x_i)$ is the likelihood that amino acid y_j will be produced if triplet x_i has been sent to the cell.

Bayes's theorem allows us to express the conditional probability $p(y_j|x_i)$ as

$$p(y_j|x_i) = \frac{p(x_i|y_j)p(y_j)}{p(x_i)} \tag{2.8}$$

and rewrite $H(Y|X)$ as

$$H(Y|X) = -\sum_{i=1}^{M}\sum_{j=1}^{L} p(x_i|y_j)p(y_j) \log \left[\frac{p(x_i|y_j)p(y_j)}{p(x_i)}\right] \tag{2.9}$$

Expanding the logarithm into two terms as

$$\log \left[\frac{p(x_i|y_j)p(y_j)}{p(x_i)}\right] = \log p(x_i|y_j) + \log \frac{p(y_j)}{p(x_i)} \tag{2.10}$$

and observing that

$$H(X|Y) = -\sum_{i=1}^{M}\sum_{j=1}^{L} p(x_i|y_j)p(y_j) \log p(x_i|y_j) \tag{2.11}$$

by definition, we obtain

$$H(Y|X) = H(X|Y) - \sum_{i=1}^{M}\sum_{j=1}^{L} p(x_i|y_j)p(y_j) \log \frac{p(y_j)}{p(x_i)} \tag{2.12}$$

2.3. REDUNDANCY AND "GENETIC NOISE"

It is usually convenient to rewrite the second term in (2.12) as

$$-\sum_{i=1}^{M}\sum_{j=1}^{L} p(x_i|y_j)p(y_j) \log \frac{p(y_j)}{p(x_i)} = -\sum_{i=1}^{M}\sum_{j=1}^{L} p(x_i)p(y_j|x_i) \log \frac{p(y_j)}{p(x_i)} \quad (2.13)$$

(using Bayes's theorem once again), which finally produces

$$H(Y|X) = H(X|Y) - \sum_{i=1}^{M}\sum_{j=1}^{L} p(x_i)p(y_j|x_i) \log \frac{p(y_j)}{p(x_i)} \quad (2.14)$$

and therefore

$$H(X|Y) = H(Y|X) + \sum_{i=1}^{M}\sum_{j=1}^{L} p(x_i)p(y_j|x_i) \log \frac{p(y_j)}{p(x_i)} \quad (2.15)$$

Replacing $H(X|Y)$ by (2.15) in expression (2.6), we now have that

$$I(X|Y) = H(X) - H(Y|X) - \sum_{i=1}^{M}\sum_{j=1}^{L} p(x_i)p(y_j|x_i) \log \frac{p(y_j)}{p(x_i)} \quad (2.16)$$

Expressions (2.15) and (2.16) are useful because they allow us to separate the loss of genetic information into two parts - a part that is due to *mutations*, and a part that is due to *redundancy*. To see this a bit more clearly, let us first recall that the term $H(Y|X)$ in (2.15) represents the remaining uncertainty about Y given that X has been sent. In the absence of genetic noise, $H(Y|X)$ would be zero, since we would know *exactly* what protein will be produced when a particular triplet x_i is sent. Under such circumstances, $H(X|Y)$ becomes

$$H(X|Y) = \sum_{i=1}^{M}\sum_{j=1}^{L} p(x_i)p(y_j|x_i) \log \frac{p(y_j)}{p(x_i)} \quad (2.17)$$

and $I(X|Y)$ reduces to

$$I(X|Y) = H(X) - \sum_{i=1}^{M}\sum_{j=1}^{L} p(x_i)p(y_j|x_i) \log \frac{p(y_j)}{p(x_i)} \quad (2.18)$$

The fact that $H(X|Y)$ remains positive even if the original "message" is not corrupted in any way suggests that a certain amount of information is irretrievably lost in the transmission process. This is due to the fact that multiple triplets of nucleotides can correspond to the *same* amino acid. Because of this intrinsic redundancy, knowing Y can never give us complete information about X, and $I(X|Y)$ cannot attain its maximal value (which is $H(X)$).

In order to compute the mutual entropy that is associated with the transmission of genetic information, we need to find a way to calculate the probabilities that figure in expression (2.16). In the following, we will focus our attention on the conditional probabilities $p(y_j|x_i)$, since $p(x_i)$ and $p(y_j)$ can be estimated from the relative frequencies with which different codons and amino acids occur in nature. To get a sense for how these probabilities can be computed, let us assume that y_j corresponds to the amino acid Leucine (*leu*), which can be produced by six different codons: CUG, CUC, CUU, CUA, UUA and UUG. What is the probability $p(y_j|x_i)$ that *leu* (which we associate with y_j) will arise if x_i corresponds to triplet UUA?

This probability would obviously be 1 in the absence of "genetic noise", since UUA is one of the codons that produces Leucine. In a realistic model, however, we must take into account the possibility of that one or more characters in the codon might be modified in the transmission process. The simplest way to incorporate this into the model would be to assume that the probability of a single mutation is α, and that this number is small enough so that mutations where two or more characters are modified are highly unlikely (which is equivalent to saying that α^2 and α^3 are negligible). Such an assumption is perfectly reasonable, and has the added benefit of making the calculations simpler.

Let us now examine what happens when only one of the characters in codon UUA is changed. If the first character is changed to C, we obtain triplet CUA, which still corresponds to Leucine. As a result, this particular mutation has no adverse effect, but the other two possibilities do, since they produce AUA (which codes for Isoleucine) and GUA (which codes for Valine). We can therefore conclude that 2 of the 3 possible modifications in the first position give rise to a different amino acid.

Any change in the second character will invariably lead to a different amino acid, since all six codons corresponding to Leucine have U in the second position. In the third position we can afford to change A into G (since UUG codes for Leucine as well), but UUU and UUC do not. We can therefore conclude that 7 of the 9 possible mutations in codon UUA alter the amino acid that is produced. Since each of these mutations has probability α, it follows that $p(y_j|x_i) = 1 - 7\alpha$.

For the sake of completeness, it might be instructive to consider one more scenario, in which x_i still corresponds to UUA, but y_j is now associated with Serine (which is produced by codons UCG, UCC, UCU and UCA). It is not difficult to see that changing U into C in the second position is the only way to produce one of these four combinations. As a result, we can conclude that in this case $p(y_j|x_i) = \alpha$.

When all possible probabilities $p(y_j|x_i)$ are computed in this manner, we have everything that we need to express $I(X|Y)$ as a function of α (note that these probabilities also allow us to compute $H(Y|X)$, using equation (1.45)).

2.3. REDUNDANCY AND "GENETIC NOISE"

An example of such a calculation can be found in Hubert Yockey's book *Information Theory, Evolution and the Origins of Life*,[7] where the author approximated the logarithm with its Taylor series expansion, and retained only terms that are linear or quadratic functions of α. In the resulting expression

$$I(X|Y) = H(X) - 1.7915 - 9.815\alpha + 34.2108\alpha^2 + 6.8303\alpha \log_2 \alpha \quad (2.19)$$

the terms that contain α can be associated with "genetic noise" (since they vanish when there are no mutations). The constant term 1.7915 that remains can then be interpreted as the *loss of genetic information due to redundancy*.[8]

Chapter 3

Quantum Information

The notion of information is not limited to communication systems and living organisms, and arises in the context of quantum mechanics as well. Quantum information "resides" in elementary particles, and is altered whenever two or more of them interact. What makes this type of information particularly interesting is the possibility of *quantum computing*, which has some significant advantages over more traditional forms of data processing.

Quantum computing is an exciting new field that brings together several different disciplines, and promises to greatly expand the scope of problems that we may be able to solve in the future. This new computing paradigm relies on the fact that the physical characteristics of elementary particles (such as momentum, spin or energy, for example) can be "encoded" in binary form. That allows us to interpret any change in the state of such a particle as a form of "computation", even though it is the result of purely natural processes.

Some scientists have taken this line of reasoning a step further, and have hypothesized that the universe itself acts like a giant "quantum computer", which uses atoms and subatomic particles as its basic "hardware".[9] Those who advocate this approach point out that it offers a plausible explanation for how an initially simple universe could give rise to the kind of complexity that we see today. This in an intriguing possibility, since it is by no means clear how random quantum fluctuations in the early universe could produce any kind of order at all (let alone the self-reproducing organic compounds that are needed to sustain life).

An analogy that is often used to explain the connection between quantum computing and the emergence of order involves a thought experiment in which monkeys are provided with typewriters, and are allowed to play with them. Given that they have no idea what they are doing, what are the odds that one of them will produce the first 100 characters of the U.S. constitution? If we assume that a typical typewriter has 50 keys, then the probability that a single monkey will generate this particular sequence of symbols is 50^{-100}, which is a fantastically small number. One would, therefore, be compelled to conclude that such an outcome is extremely improbable unless some additional requirements are imposed.

To see what kinds of requirements would be appropriate in this case, let us imagine a scenario in which the monkeys are now allowed to type random sequences of characters *into a computer* rather than a typewriter. What would change? Quite a bit, actually, since we know that short randomly generated programs can sometimes produce very elaborate and complex outputs. The lesson that we can draw from this is that randomness can potentially play a constructive role in the domain of computing. Such a conclusion obviously fits very nicely with a model in which the universe acts as a quantum computer, and quantum fluctuations provide random "programs" for it to execute.

It goes without saying, of course, that all this is still highly speculative, and that we are in no position to tell whether or not the universe actually operates in this manner. What we do know, however, is that quantum information exists, and that it can be meaningfully manipulated by devices known as *quantum gates*. Systems that use these gates as their basic "building blocks" differ significantly from classical computers, since they have *both* digital *and* analog aspects. This feature provides an additional degree of freedom, which can be exploited to solve certain problems that are currently considered to be exceptionally difficult.

An obvious example of this sort is prime factorization, which can be performed efficiently using quantum computers although there is no comparable algorithm for digital computers.[10] Something similar can be said about simulating quantum systems that consist of N entangled particles. This type of computation poses a major challenge because the wave function that describes such a system consists of 2^N terms, each of which requires a certain amount of storage space. If we tried to manipulate all this data using a conventional computer, we would easily run out of memory.

When making such comparisons, it is important to keep in mind that quantum computing is not "omnipotent", and has certain limitations of its own. One of the most important ones stems from the fact that quantum information processing requires a series of changes in the states of elementary particles. Since the maximum rate of change is proportional to the amount of available energy, it follows that there is an upper bound on the speed with which a quantum computer can manipulate and transmit information. This bound can be estimated using the Margolus-Levitin theorem,[11] which tells us that an atom can change its state up to 3×10^{13} times per second. That is about four orders of magnitude faster than the maximal rate at which conventional computers can operate.[12]

In the sections that follow, we will explore some of these questions in greater detail, and outline the basic ideas and results that make quantum computing possible. We will also examine the functional characteristics of different types of quantum gates, and explain how such devices differ from standard logic circuits. As we do so, we will make a concerted effort to highlight certain features that are unique to quantum computers, and cannot be replicated by their digital counterparts.

3.1 An Overview of Quantum Mechanics

The state of a quantum particle is described by a complex-valued *wave function*, which is usually denoted by $\psi(x,t)$. The evolution of this function over time is governed by Schrödinger's equation

$$i\hbar \frac{\partial \psi}{\partial t} = \hat{H}\psi \tag{3.1}$$

where $\hat{H}$ represents a linear operator, and $\hbar$ is Planck's constant divided by 2π. For a single particle of mass m in an external field $V(x,y,z)$, $\hat{H}$ has the form

$$\hat{H} = -\frac{\hbar^2}{2m}\nabla^2 + V(x,y,z) \tag{3.2}$$

where

$$\nabla^2 = \frac{\partial^2}{\partial x^2} + \frac{\partial^2}{\partial y^2} + \frac{\partial^2}{\partial z^2} \tag{3.3}$$

Although $\hat{H}$ is an abstract mathematical operator with no obvious physical significance, it has some important similarities with the total energy of a classical particle. To see this more clearly, we should recall that in the classical case the sum of the kinetic and potential energy can be expressed as

$$E = \frac{1}{2}m(v_x^2 + v_y^2 + v_z^2) + V(x,y,z) \tag{3.4}$$

or equivalently (in terms of momentum) as

$$E = \frac{1}{2m}(p_x^2 + p_y^2 + p_z^2) + V(x,y,z) \tag{3.5}$$

If the momentum components in (3.5) are now replaced by *momentum operators*

$$\hat{p}_x = i\hbar\frac{\partial}{\partial x}; \quad \hat{p}_y = i\hbar\frac{\partial}{\partial y}; \quad \hat{p}_z = i\hbar\frac{\partial}{\partial z} \tag{3.6}$$

we obtain

$$\frac{1}{2m}(\hat{p}_x^2 + \hat{p}_y^2 + \hat{p}_z^2) = -\frac{\hbar^2}{2m}\nabla^2 \tag{3.7}$$

which matches the first term in $\hat{H}$ exactly.

What does the wave function $\psi(x,t)$ actually tell us? That depends to some extent on what it is that we want to know. Suppose, for example, that we are interested in measuring some observable quantity q. With any such quantity, we can associate an operator $\hat{Q}$, much like we did with the momentum operator in equation (3.6). By definition, the eigenfunctions of $\hat{Q}$ (which are denoted $\phi_i(x)$) satisfy

$$\hat{Q}\phi_i(x) = q_i\phi_i(x) \tag{3.8}$$

where q_i ($i = 1, 2, \ldots$) represent the corresponding *eigenvalues*. The quantum mechanical interpretation of this expression is that a particle is in a *definite state of q* only when $\psi(x)$ equals one of the eigenfunctions $\{\phi_1(x), \phi_2(x), \ldots\}$. This is equivalent to saying that a measurement is guaranteed to produce $q = q_i$ if and only if $\psi(x) = \phi_i(x)$.

To get a sense for how quantum mechanics handles states that do *not* correspond to eigenfunctions, we should first note that operator $\hat{Q}$ is always chosen so that it has the following three properties:

Property 1. All eigenvalues of $\hat{Q}$ must be *real numbers*. This is important because it allows us to identify observable physical quantities with eigenvalues.

Property 2. The eigenfunctions of $\hat{Q}$ form a *basis*, which implies that *any* function $f(x)$ can be expressed in terms of $\{\phi_1(x), \phi_2(x), \ldots\}$ as

$$f(x) = \sum_i \beta_i \phi_i(x) \tag{3.9}$$

where β_i ($i = 1, 2, \ldots$) are constant coefficients.

Property 3. The eigenfunctions of $\hat{Q}$ are *orthonormal*. This means that their scalar product

$$\langle \phi_i(x), \phi_j(x) \rangle = \int \phi_i^*(\rho)\phi_j(\rho)d\rho \tag{3.10}$$

satisfies

$$\langle \phi_i(x), \phi_j(x) \rangle = \begin{cases} 1, & i = j \\ 0, & i \neq j \end{cases} \tag{3.11}$$

When $\psi(x)$ is *not* an eigenfunction of $\hat{Q}$, Property 2 allows us to express it as

$$\psi(x) = \sum_i \alpha_i(q_i)\phi_i(x) \tag{3.12}$$

In such cases, we say that the particle is in a *state of superposition* (as opposed to a state of definite q). What this means is that a measurement could produce

3.1. AN OVERVIEW OF QUANTUM MECHANICS

any of the possible values for q, but we cannot tell in advance which one. All we can possibly know is the probability of measuring a particular value q_i, which can be calculated as

$$P(q_i) = |\alpha_i(q_i)|^2 \tag{3.13}$$

where $\alpha_i(q_i)$ are the coefficients that appear in (3.12).

To see how $P(q_i)$ depends on functions $\psi(x)$ and $\phi_i(x)$, we should recall that the scalar product defined in equation (3.10) satisfies

$$\langle \sum_i a_i f_i(x), g(x) \rangle = \sum_i a_i^* \langle f_i(x), g(x) \rangle \tag{3.14}$$

for any two complex-valued functions $f_i(x)$ and $g(x)$. Using this property in conjunction with equations (3.11) and (3.12), we obtain

$$\langle \psi(x), \phi_i(x) \rangle = \sum_j \alpha_j^*(q_j) \langle \phi_j(x), \phi_i(x) \rangle = \alpha_i^*(q_i) \tag{3.15}$$

which allows us to express $P(q_i)$

$$P(q_i) = |\alpha_i(q_i)|^2 = |\langle \psi(x), \phi_i(x) \rangle|^2. \tag{3.16}$$

This relationship provides us with a very simple way to calculate the probabilities of all possible outcomes.

Remark 3.1. In order to treat the terms $|\alpha_i(q_i)|^2$ as probabilities, we must ensure that

$$\sum_i |\alpha_i(q_i)|^2 = 1 \tag{3.17}$$

We can do this by requiring that $\psi(x)$ satisfies $\langle \psi(x), \psi(x) \rangle = 1$ (in which case we say that it is *normalized*). To see why normalization is equivalent to condition (3.17), it suffices to observe that

$$\langle \psi(x), \psi(x) \rangle = \langle \sum_i \alpha_i \phi_i(x), \sum_j \alpha_j \phi_j(x) \rangle =$$
$$= \sum_{i,j} \alpha_i^* \alpha_j \langle \phi_i(x), \phi_j(x) \rangle = \sum_i |\alpha_i(q_i)|^2 \tag{3.18}$$

What happens when we perform a measurement on a particle that is in a state of superposition? According to the so-called Copenhagen interpretation of quantum mechanics, its wave function *collapses* into exactly one of the eigenfunctions $\phi_i(x)$, and automatically moves into a state that corresponds to eigenvalue q_i. After the initial observation is made, all subsequent tests will invariably produce the same result, since the system remains in state $\psi(x) = \phi_i(x)$. In this state, the probability of detecting some $q_j \neq q_i$ is

$$P(q_j) = |\langle \phi_i(x), \phi_j(x) \rangle|^2 = 0 \tag{3.19}$$

Quantum Entanglement

When two or more quantum particles interact, their wave functions become "entangled" in a way that does not allow us to reduce the behavior of the system to the behavior of its constituent parts. In order to understand what this means, let us first consider a classical system S, which consists of two subsystems S_1 and S_2. If the possible states of S_1 and S_2 are $\{x_1, x_2, x_3, \ldots\}$ and $\{y_1, y_2, y_3, \ldots\}$, respectively, the state of the overall system can always be uniquely associated with a pair of "microstates" (x_i, y_j). This allows us to describe the behavior of the "whole" in terms of the behavior of its constituent parts.

In quantum mechanics, the situation is considerably more complicated. To see why this is so, let us consider a simple two particle system whose constituents are described by wave functions ψ and φ. We will assume that ψ belongs to some functional space H_1 whose basis $\{\psi_1, \psi_2, \psi_3, \ldots\}$ is associated with measurable values of some physical quantity a. We will make a similar assumption for system S_2 as well, the only difference being that its state φ belongs to functional space H_2 whose basis $\{\varphi_1, \varphi_2, \varphi_3, \ldots\}$ is associated with measurable values of some physical quantity b.

When these two subsystems *interact*, the state of the resulting composite system S can be described as

$$\chi = \sum_{i,\,j} \alpha_{ij}(\psi_i \otimes \varphi_j) \tag{3.20}$$

where $\otimes$ denotes a mathematical operation known as the *tensor product*. This product effectively combines the functional spaces H_1 and H_2 into a new space $H = H_1 \otimes H_2$. An important feature of this space is that it contains elements that cannot be represented as $\psi \otimes \varphi$, where $\psi \in H_1$ and $\varphi \in H_2$. As a result, certain states of the composite system *cannot be fully described in terms of the states of the original subsystems* (which is not the case in classical systems).

Without getting into further details, it will suffice to note that the tensor product has three basic properties:

1. If $\{\psi_1, \psi_2, \psi_3, \ldots\}$ and $\{\varphi_1, \varphi_2, \varphi_3, \ldots\}$ are bases in spaces H_1 and H_2, then any vector in space $H = H_1 \otimes H_2$ can be represented in the manner shown in (3.20).

2. If functions ψ, φ and ξ belong to spaces H_1, H_2 and H_3, respectively, and α and β are complex numbers, the following three identities hold true:

$$(\alpha\psi + \beta\varphi) \otimes \xi = (\alpha\psi) \otimes \xi + (\beta\varphi) \otimes \xi \tag{3.21}$$

$$\psi \otimes (\alpha\varphi + \beta\xi) = \psi \otimes (\alpha\varphi) + \psi \otimes (\beta\xi) \tag{3.22}$$

$$\alpha(\psi \otimes \varphi) = (\alpha\psi) \otimes \varphi = \psi \otimes (\alpha\varphi) \tag{3.23}$$

3. Let $\hat{A}_1$ and $\hat{A}_2$ be linear operators in spaces H_1 and H_2, respectively. We can then define operator $\hat{A}_1 \otimes \hat{A}_2$ in space $H = H_1 \otimes H_2$ as

$$\left(\hat{A}_1 \otimes \hat{A}_2\right)(\psi \otimes \varphi) = (\hat{A}_1\psi) \otimes (\hat{A}_2\varphi) \tag{3.24}$$

Since operator $\hat{A}_1 \otimes \hat{A}_2$ is assumed to be linear, applying it to a composite function such as the one in (3.20) produces

$$\left(\hat{A}_1 \otimes \hat{A}_2\right)\chi = \sum_{i,j} \alpha_{ij}(\hat{A}_1\psi_i) \otimes (\hat{A}_2\varphi_j) \tag{3.25}$$

3.2 Quantum Computing

In contrast to traditional computing (which operates on *bits* of information), quantum computing works with *qubits*. What exactly does this term mean? Qubits can be identified with quantum states of the form $\psi = \alpha\psi_0 + \beta\psi_1$, where ψ_0 and ψ_1 correspond to observable values of some physical quantity. The specific interpretation of states ψ_0 and ψ_1 depends on the physical realization of the system. It can refer, for example, to the spin of an atom (in which case 0 and 1 correspond to "up" and "down" orientations along an axis), to its energy state (where 0 and 1 represent the "ground" and the "excited" state, respectively), or to any number of other physical properties with binary characteristics. Although such a definition directly associates qubits with physical properties of quantum particles, for our purposes it will be more convenient to think of a qubit as an *abstract mathematical entity*. The advantage of such an interpretation is that it allows us to develop a theory of quantum computing that is *independent* of the specific implementation.

One obvious difference between qubits and ordinary bits stems from the fact that the state of a qubit is *not* limited to ψ_0 and ψ_1. In addition to these two states (which correspond to the "classical" 0 and 1), a qubit can also be in a *state of superposition*, in which both α and β are nonzero. There is an unlimited number of such "intermediate" states, since the only constraint that these coefficients must satisfy is $|\alpha|^2 + |\beta|^2 = 1$. It is important to note, however, that the information contained in the state of superposition is "hidden", since any measurement performed on such a system will necessarily produce either ψ_0 or ψ_1 (with probabilities $|\alpha|^2$ and $|\beta|^2$, respectively). These two states tell us nothing about what the function looked like before it collapsed - it could have been in *any* state $\psi = \alpha\psi_0 + \beta\psi_1$.

How do quantum computers exploit this "hidden" information? To see that, let us consider a 2-qubit system which consists of a pair of identical particles a and b. If we assume that each particle has the same measurable states ψ_0 and ψ_1, we have that

$$\psi_a = \alpha_1 \psi_0 + \beta_1 \psi_1 \tag{3.26}$$

and

$$\psi_b = \alpha_2 \psi_0 + \beta_2 \psi_1 \tag{3.27}$$

After the two particles interact, their wave functions become entangled, and their composite state becomes

$$\Psi = \psi_a \otimes \psi_b = (\alpha_1 \psi_0 + \beta_1 \psi_1) \otimes (\alpha_2 \psi_0 + \beta_2 \psi_1) \tag{3.28}$$

Expanding this expressions according to the tensor product rules described in equations (3.21)-(3.23), we obtain

$$\Psi = \alpha_1 \alpha_2 (\psi_0 \otimes \psi_0) + \alpha_1 \beta_2 (\psi_0 \otimes \psi_1) + \beta_1 \alpha_2 (\psi_1 \otimes \psi_0) + \\ + \beta_1 \beta_2 (\psi_1 \otimes \psi_1) = a_0 \Psi_{00} + a_1 \Psi_{01} + a_2 \Psi_{10} + a_3 \Psi_{11} \tag{3.29}$$

where Ψ_{ij} denotes the tensor product $\psi_i \otimes \psi_j$. Recalling that

$$|\alpha_1|^2 + |\beta_1|^2 = |\alpha_2|^2 + |\beta_2|^2 = 1 \tag{3.30}$$

and observing that $|a_0|^2 = |\alpha_1|^2 |\alpha_2|^2$ and $|a_1|^2 = |\alpha_1|^2 |\beta_2|^2$ it follows that

$$|a_0|^2 + |a_1|^2 = |\alpha_1|^2 \left[|\alpha_2|^2 + |\beta_2|^2 \right] = |\alpha_1|^2 \tag{3.31}$$

Similarly, from $|a_2|^2 = |\beta_1|^2 |\alpha_2|^2$ and $|a_3|^2 = |\beta_1|^2 |\beta_2|^2$ we get

$$|a_2|^2 + |a_3|^2 = |\beta_1|^2 \left[|\alpha_2|^2 + |\beta_2|^2 \right] = |\beta_1|^2 \tag{3.32}$$

This tells us that the aggregate wave function Ψ of the 2-qubit system is *normalized*, since

$$|a_0|^2 + |a_1|^2 + |a_2|^2 + |a_3|^2 = 1 \tag{3.33}$$

What happens if we now decide to perform a measurement only on the *first* qubit (while leaving the second one in a state of superposition)? The probability that we will record, say, a 0 is $|a_0|^2 + |a_1|^2$, since *both* Ψ_{00} and Ψ_{01} correspond to such an outcome. Note, however, that at this point we still don't know what the state of the other particle will be when we measure it. The pair will therefore remain in a state of superposition, which is characterized by the wave function

$$\Psi = b_0 \Psi_{00} + b_1 \Psi_{01} \tag{3.34}$$

3.2. QUANTUM COMPUTING

where

$$b_0 = \frac{a_0}{\sqrt{a_0^2 + a_1^2}}; \quad \text{and} \quad b_1 = \frac{a_1}{\sqrt{a_0^2 + a_1^2}} \quad (3.35)$$

represent *renormalized coefficients*. This wave function could collapse either into state Ψ_{00} or Ψ_{01}, depending on the results of the measurement made on the second particle.

The ideas outlined above can be easily extended to systems of n identical particles, in which case Ψ has the general form

$$\Psi = a_0 \Psi_{00\ldots0} + a_1 \Psi_{00\ldots1} + \ldots + a_{2^n-1} \Psi_{11\ldots1} \quad (3.36)$$

and the coefficients a_i satisfy

$$\sum_{i=1}^{2^n} |a_i|^2 = 1 \quad (3.37)$$

This suggests that the number of measurable states grows *exponentially* as the system size increases. To see why this is important, imagine a system that consists of 300 particles (which is by no means unrealistic). The number of possible outcomes in such a system would be a staggering $2^{300} \approx 2 \times 10^{90}$, which exceeds the estimated number of particles in the universe.

3.2.1 Single Qubit Gates

Understanding how a quantum computer works requires at least a cursory familiarity with its basic "building blocks", which are known as *quantum gates*. As the name suggests, quantum gates have certain similarities with standard logic gates that are used in the design of digital computers. There are, however, some fundamental differences as well, one of which has to do with the fact that quantum gates operate on the *coefficients that describe the state of superposition* (rather than the measurable states themselves).

To see why this is an advantage, we should note that a standard logic gate with n inputs can perform only *one* operation at a time (an n-input AND gate, for example, transforms any input combination $(X_1, X_2, \ldots, X_n)$ into $Z = X_1 \cdot X_2 \cdot \ldots \cdot X_n$). An n-input quantum gate, on the other hand, operates on all 2^n coefficients in expression (3.36) *simultaneously*, and transforms them into a different set of coefficients $\{b_0, b_1, \ldots, b_{2^n-1}\}$. In the sections that follow we will show how this works in practice, starting with basic quantum gates that operate on a single qubit.

The Quantum NOT Gate

Given a single qubit in state $\psi = \alpha \psi_0 + \beta \psi_1$, the quantum NOT operation transforms it into

$$\phi = \hat{X} \psi = \beta \psi_0 + \alpha \psi_1 \quad (3.38)$$

where coefficients α and β *switch places*. If we express function ϕ as

$$\phi = \hat{X}\psi = \rho_1\psi_0 + \rho_2\psi_1 \qquad (3.39)$$

we can say that the coefficients ρ_1 and ρ_2 are related to α and β as

$$\begin{bmatrix} \rho_1 \\ \rho_2 \end{bmatrix} = \begin{bmatrix} 0 & 1 \\ 1 & 0 \end{bmatrix} \begin{bmatrix} \alpha \\ \beta \end{bmatrix} \qquad (3.40)$$

It is not difficult to see that repeating this operation will return the qubit to its original state, since

$$\begin{bmatrix} 0 & 1 \\ 1 & 0 \end{bmatrix} \begin{bmatrix} \rho_1 \\ \rho_2 \end{bmatrix} = \begin{bmatrix} 0 & 1 \\ 1 & 0 \end{bmatrix} \begin{bmatrix} 0 & 1 \\ 1 & 0 \end{bmatrix} \begin{bmatrix} \alpha \\ \beta \end{bmatrix} = \begin{bmatrix} \alpha \\ \beta \end{bmatrix} \qquad (3.41)$$

We can therefore conclude that operator $\hat{X}$ satisfies $\hat{X}^2 = I$, and that the corresponding transformation $\phi = \hat{X}\psi$ is reversible (an operator with these properties is said to be *unitary*).

It is important to recognize that the quantum NOT gate is more general than its classical counterpart, since it *simultaneously* manipulates both α and β (rather than a single 0 or 1). The two operations match only if $(\alpha = 1, \beta = 0)$ or $(\alpha = 0, \beta = 1)$, in which case the quantum transformations

$$\hat{X}\psi_0 = 0 \cdot \psi_0 + 1 \cdot \psi_1 = \psi_1 \qquad (3.42)$$

and

$$\hat{X}\psi_1 = 1 \cdot \psi_0 + 0 \cdot \psi_1 = \psi_0 \qquad (3.43)$$

reduce to simple "bit flips" (of the sort that a standard NOT gate performs). This is an important difference, since it implies that quantum gates can produce an unlimited number of internal states *beyond* the ones that are actually measurable. Although we cannot observe them, these states represent an essential part of the computation.

The Z - Gate

Given a single qubit in state $\psi = \alpha\psi_0 + \beta\psi_1$, the Z-gate transforms it into state

$$\phi = \hat{Z}\psi = \alpha\psi_0 - \beta\psi_1 \qquad (3.44)$$

If we once again express function ϕ as

$$\phi = \hat{Z}\psi = \rho_1\psi_0 + \rho_2\psi_1 \qquad (3.45)$$

we have that

$$\begin{bmatrix} \rho_1 \\ \rho_2 \end{bmatrix} = \begin{bmatrix} 1 & 0 \\ 0 & -1 \end{bmatrix} \begin{bmatrix} \alpha \\ \beta \end{bmatrix} \qquad (3.46)$$

3.2. QUANTUM COMPUTING

As in the case of the quantum NOT gate, repeating this operation will return the qubit to its original state, which implies that $\hat{Z}$ is a *unitary operator*.

The Hadamard Gate

A third type of single qubit gate is the so-called Hadamard gate, which transforms the state $\psi = \alpha\psi_0 + \beta\psi_1$ into

$$\phi = \hat{H}\psi = \rho_1\psi_0 + \rho_2\psi_1 \tag{3.47}$$

where

$$\rho_1 = \frac{1}{\sqrt{2}}(\alpha + \beta) \quad \text{and} \quad \rho_2 = \frac{1}{\sqrt{2}}(\alpha - \beta) \tag{3.48}$$

In matrix form, this relationship can be described as

$$\begin{bmatrix} \rho_1 \\ \rho_2 \end{bmatrix} = \frac{1}{\sqrt{2}} \begin{bmatrix} 1 & 1 \\ 1 & -1 \end{bmatrix} \begin{bmatrix} \alpha \\ \beta \end{bmatrix} \tag{3.49}$$

Repeating such an operation produces

$$\frac{1}{\sqrt{2}} \begin{bmatrix} 1 & 1 \\ 1 & -1 \end{bmatrix} \begin{bmatrix} \rho_1 \\ \rho_2 \end{bmatrix} = \frac{1}{2} \begin{bmatrix} 1 & 1 \\ 1 & -1 \end{bmatrix} \begin{bmatrix} 1 & 1 \\ 1 & -1 \end{bmatrix} \begin{bmatrix} \alpha \\ \beta \end{bmatrix} = \begin{bmatrix} \alpha \\ \beta \end{bmatrix} \tag{3.50}$$

which implies that $\hat{H}$, too, is a *unitary operator* (just like $\hat{Z}$ and $\hat{X}$).

The Hadamard operation

$$\phi = \hat{H}\psi = \frac{1}{\sqrt{2}}(\alpha + \beta)\psi_0 + \frac{1}{\sqrt{2}}(\alpha - \beta)\psi_1 \tag{3.51}$$

can be equivalently rewritten as

$$\phi = \alpha\left(\frac{\psi_0 + \psi_1}{\sqrt{2}}\right) + \beta\left(\frac{\psi_0 - \psi_1}{\sqrt{2}}\right) = \alpha\psi_+ + \beta\psi_- \tag{3.52}$$

It is not difficult to show that functions ψ_+ and ψ_- form an *orthonormal basis* in their own right, which means that they correspond to a pair of measurable physical states. These states are different from ψ_0 and ψ_1, but the probabilities of observing them are the same (they are equal to $|\alpha|^2$ and $|\beta|^2$, respectively).

In the special case when ψ is equal to ψ_0 or ψ_1, the Hadamard operation produces $\hat{H}\psi_0 = \psi_+$ and $\hat{H}\psi_1 = \psi_-$. With that in mind, it would be interesting to consider what happens if we feed ψ_0 to *two* such gates, and entangle the outputs. In that case, we have

$$\psi_+ \otimes \psi_+ = \frac{1}{2}[(\psi_0 + \psi_1) \otimes (\psi_0 + \psi_1)] = \frac{1}{2}[\Psi_{00} + \Psi_{01} + \Psi_{10} + \Psi_{11}] \tag{3.53}$$

where all four outcomes have the *same* probability.

The operation described in equation (3.53) is usually denoted $H^{\otimes 2}(\psi_0)$, and plays an important role in the design of quantum algorithms. If we perform it on *three* qubits that were initially in state ψ_0 (which corresponds to $H^{\otimes 3}(\psi_0)$), we obtain

$$\psi_+ \otimes (\psi_+ \otimes \psi_+) = \frac{1}{\sqrt{2}}(\psi_0 + \psi_1) \otimes \frac{1}{2}(\Psi_{00} + \Psi_{01} + \Psi_{10} + \Psi_{11}) \qquad (3.54)$$

Denoting

$$\psi_i \otimes \Psi_{jk} = \psi_i \otimes (\psi_j \otimes \psi_k) \equiv \Psi_{ijk} \qquad (3.55)$$

equation (3.54) becomes

$$\psi_+ \otimes (\psi_+ \otimes \psi_+) = \frac{1}{\sqrt{2^3}}(\Psi_{000} + \Psi_{001} + \ldots + \Psi_{110} + \Psi_{111}) \qquad (3.56)$$

This procedure can obviously be generalized to include n Hadamard gates (each with ψ_0 as its input), in which case we obtain a composite wave function with 2^n components. This function can be represented in compact form as

$$H^{\otimes n}(\psi_0) = \frac{1}{\sqrt{2^n}} \sum_{x=0}^{2^n - 1} \hat{\Psi}_x \qquad (3.57)$$

where $\hat{\Psi}_0 = \Psi_{00\ldots 0}$, $\hat{\Psi}_1 = \Psi_{00\ldots 1}$, ..., $\hat{\Psi}_{2^n - 1} = \Psi_{11\ldots 1}$. Note that $H^{\otimes n}(\psi_0)$ has a similar form to the one in expression (3.36), except for the fact that states $\hat{\Psi}_x$ now have the *same* probability

$$p = \left(\frac{1}{\sqrt{2^n}}\right)^2 = \frac{1}{2^n} \qquad (3.58)$$

3.2.2 The C-NOT Gate

Perhaps the most important multiple qubit gate is the so-called C-NOT gate, whose schematic diagram is shown in Fig. 3.1. This type of gate is said to be *universal* because it can be used to construct *any* other multiple qubit gate.

3.2. QUANTUM COMPUTING

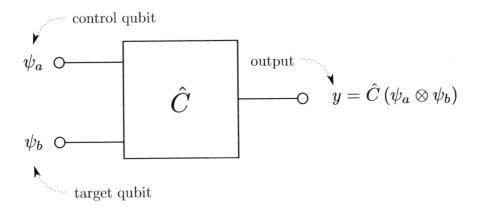

Figure 3.1: Schematic representation of a C-NOT gate.

To see how a C-NOT gate operates, let us assume that the *control qubit* is in state $\psi_a = \alpha\psi_0 + \beta\psi_1$ and that the *target qubit* is in state $\psi_b = \psi_0$. Once the two input qubits become entangled, the operator $\hat{C}$ acts on the resulting function

$$\psi_a \otimes \psi_0 = \alpha(\psi_0 \otimes \psi_0) + \beta(\psi_1 \otimes \psi_0) = \alpha\Psi_{00} + \beta\Psi_{10} \tag{3.59}$$

in the following way:

$$\hat{C}(\psi_a \otimes \psi_0) = \alpha\hat{C}(\psi_0 \otimes \psi_0) + \beta\hat{C}(\psi_1 \otimes \psi_0) =$$
$$= \alpha(\psi_0 \otimes \psi_0) + \beta(\psi_1 \otimes \psi_1) = \alpha\Psi_{00} + \beta\Psi_{11} \tag{3.60}$$

In other words, the second index in function Ψ_{ij} is "flipped" if the first index is 1, and remains intact otherwise.

In a more general scenario, the target bit could be in a state of superposition $\psi_b = \gamma\psi_0 + \xi\psi_1$, in which case $\hat{C}$ operates on function

$$\psi_a \otimes \psi_b = (\alpha\psi_0 + \beta\psi_1) \otimes (\gamma\psi_0 + \xi\psi_1) =$$
$$\gamma\alpha(\psi_0 \otimes \psi_0) + \xi\alpha(\psi_0 \otimes \psi_1) + \gamma\beta(\psi_1 \otimes \psi_0) + \xi\beta(\psi_1 \otimes \psi_1) \tag{3.61}$$

Setting $a_0 = \gamma\alpha$, $a_1 = \xi\alpha$, $a_2 = \gamma\beta$ and $a_3 = \xi\beta$, this expression can be rewritten in compact form as

$$\psi_a \otimes \psi_b = a_0\Psi_{00} + a_1\Psi_{01} + a_2\Psi_{10} + a_3\Psi_{11} \tag{3.62}$$

Since operator $\hat{C}$ satisfies

$$\hat{C}(\Psi_{00}) = \Psi_{00}$$
$$\hat{C}(\Psi_{01}) = \Psi_{01}$$
$$\hat{C}(\Psi_{10}) = \Psi_{11} \tag{3.63}$$
$$\hat{C}(\Psi_{11}) = \Psi_{10}$$

we have that
$$\Phi = \hat{C}(\psi_a \otimes \psi_b) = a_0\Psi_{00} + a_1\Psi_{01} + a_2\Psi_{11} + a_3\Psi_{10} \qquad (3.64)$$

If we rewrite (3.64) as
$$\Phi = b_0\Psi_{00} + b_1\Psi_{01} + b_2\Psi_{10} + b_3\Psi_{11} \qquad (3.65)$$

(which conforms to the ordering in (3.62)), the relationship between coefficients $[a_0\ a_1\ a_2\ a_3]$ and $[b_0\ b_1\ b_2\ b_3]$ can be expressed in matrix form as

$$\begin{bmatrix} b_0 \\ b_1 \\ b_2 \\ b_3 \end{bmatrix} = \begin{bmatrix} 1 & 0 & 0 & 0 \\ 0 & 1 & 0 & 0 \\ 0 & 0 & 0 & 1 \\ 0 & 0 & 1 & 0 \end{bmatrix} \begin{bmatrix} a_0 \\ a_1 \\ a_2 \\ a_3 \end{bmatrix} \qquad (3.66)$$

where
$$C = \begin{bmatrix} 1 & 0 & 0 & 0 \\ 0 & 1 & 0 & 0 \\ 0 & 0 & 0 & 1 \\ 0 & 0 & 1 & 0 \end{bmatrix} \qquad (3.67)$$

It is not difficult to verify that $C^2 = I$, which implies that $\hat{C}$ is a unitary operator, and that the C-NOT operation is reversible.

C-NOT *Gates and Bell States*

In addition to being a universal building block for quantum computers, the C-NOT gate can also be used to produce so-called *Bell states*, which play an important role in quantum mechanics. States of this sort take one of the following four forms

$$\Phi_{00} = \frac{1}{\sqrt{2}}(\Psi_{00} + \Psi_{11}) \qquad (3.68)$$

$$\Phi_{01} = \frac{1}{\sqrt{2}}(\Psi_{01} + \Psi_{10}) \qquad (3.69)$$

$$\Phi_{10} = \frac{1}{\sqrt{2}}(\Psi_{00} - \Psi_{11}) \qquad (3.70)$$

$$\Phi_{11} = \frac{1}{\sqrt{2}}(\Psi_{01} - \Psi_{10}) \qquad (3.71)$$

and can be obtained using the configuration shown in Fig. 3.2 (which consists of a C-NOT gate and a Hadamard gate).

The Hadamard gate takes ψ_0 as its input, and generates function

$$\psi_+ = \frac{1}{\sqrt{2}}\psi_0 + \frac{1}{\sqrt{2}}\psi_1 \qquad (3.72)$$

3.2. QUANTUM COMPUTING

in the manner described in equation (3.52) (after setting $\alpha = 1$ and $\beta = 0$).

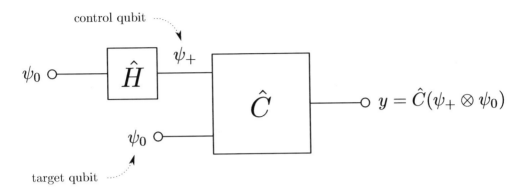

Figure 3.2: A combination of gates that produces Bell state Φ_{00}.

The C-NOT gate then uses ψ_+ as the *control qubit* and ψ_0 as the *target qubit*, producing

$$\hat{C}\left(\psi_+ \otimes \psi_0\right) = \frac{1}{\sqrt{2}}\hat{C}\left(\psi_0 \otimes \psi_0\right) + \frac{1}{\sqrt{2}}\hat{C}\left(\psi_1 \otimes \psi_0\right) =$$
$$= \frac{1}{\sqrt{2}}\hat{C}(\Psi_{00}) + \frac{1}{\sqrt{2}}\hat{C}(\Psi_{10}) = \frac{1}{\sqrt{2}}\left(\Psi_{00} + \Psi_{11}\right) \equiv \Phi_{00} \quad (3.73)$$

To realize the Bell function Φ_{01}, we simply need to replace the target qubit with ψ_1, in which case we obtain

$$\hat{C}\left(\psi_+ \otimes \psi_1\right) = \frac{1}{\sqrt{2}}\hat{C}\left(\psi_0 \otimes \psi_1\right) + \frac{1}{\sqrt{2}}\hat{C}\left(\psi_1 \otimes \psi_1\right) =$$
$$= \frac{1}{\sqrt{2}}\hat{C}(\Psi_{01}) + \frac{1}{\sqrt{2}}\hat{C}(\Psi_{11}) = \frac{1}{\sqrt{2}}\left(\Psi_{01} + \Psi_{10}\right) \equiv \Phi_{01} \quad (3.74)$$

If we use ψ_1 as the input to the Hadamard gate (instead of ψ_0), the control qubit becomes

$$\psi_- = \frac{1}{\sqrt{2}}\psi_0 - \frac{1}{\sqrt{2}}\psi_1 \quad (3.75)$$

Setting ψ_0 as the target qubit, we obtain

$$\hat{C}\left(\psi_- \otimes \psi_0\right) = \frac{1}{\sqrt{2}}\hat{C}\left(\psi_0 \otimes \psi_0\right) - \frac{1}{\sqrt{2}}\hat{C}\left(\psi_1 \otimes \psi_0\right) =$$
$$= \frac{1}{\sqrt{2}}\hat{C}(\Psi_{00}) - \frac{1}{\sqrt{2}}\hat{C}(\Psi_{10}) = \frac{1}{\sqrt{2}}\left(\Psi_{00} - \Psi_{11}\right) \equiv \Phi_{10} \quad (3.76)$$

while choosing ψ_1 as the target qubit yields

$$\begin{aligned}\hat{C}\left(\psi_{-}\otimes\psi_{1}\right) &= \frac{1}{\sqrt{2}}\hat{C}\left(\psi_{0}\otimes\psi_{1}\right)-\frac{1}{\sqrt{2}}\hat{C}\left(\psi_{1}\otimes\psi_{1}\right) = \\ &= \frac{1}{\sqrt{2}}\hat{C}(\Psi_{01})-\frac{1}{\sqrt{2}}\hat{C}(\Psi_{11}) = \frac{1}{\sqrt{2}}\left(\Psi_{01}-\Psi_{10}\right)\equiv\Phi_{11}\end{aligned} \quad (3.77)$$

The Relationship Between C-NOT *and* XOR *Gates*

The C-NOT gate is often viewed as a quantum generalization of the classical XOR gate. To see why, we should first recall that the truth table for the standard XOR operation has the form

X	Y	$W = X \oplus Y$
0	0	0
0	1	1
1	0	1
1	1	0

Table 3.1. The classical XOR operation.

If we interpret X as the "control" bit and Y as the "target" bit, then Y clearly "flips" when $X = 1$ and remains unchanged otherwise (more formally, we can say that $0 \oplus Y = Y$ and $1 \oplus Y = \bar{Y}$). This is precisely what the quantum C-NOT gate does as well, but there are several significant differences that we need to take into account. One of them has to do with the fact that an XOR gate always performs a *single* operation on inputs X and Y, producing $W = X \oplus Y$ as the output. A C-NOT gate, on the other hand, simultaneously operates on *all four* coefficients $[a_0\ a_1\ a_2\ a_3]$ and transforms them into $[b_0\ b_1\ b_2\ b_3]$. This suggests that quantum gates have an inherent potential for parallelism that no conventional logic gate can replicate.

Another important difference between C-NOT and XOR gates is that the C-NOT operation is *reversible*, while XOR is *not*. Indeed, given $W = 1$, no subsequent operation on output W will allow us to recover X and Y (since *two* different combinations correspond to the same output). Something similar can be said for the NAND gate as well, whose truth table is

X	Y	$Z = \overline{X \cdot Y}$
0	0	1
0	1	1
1	0	1
1	1	0

Table 3.2. The classical NAND operation.

3.2. QUANTUM COMPUTING

In this case, the output $W = 1$ corresponds to *three* possible input combinations, so we are once again unable to determine the original X and Y.

Yet another difference between C-NOT and XOR gates is related to a property that is sometimes referred to as the *No-Cloning Theorem*. To understand what this means, we should first observe that classical bits can always be copied in the manner illustrated in Fig. 3.3 (where $\oplus$ denotes the XOR operation).

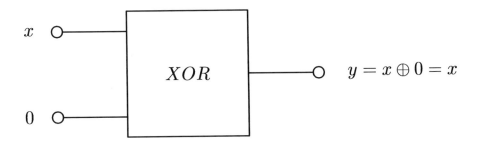

Figure 3.3: The "copy" operation using an XOR gate.

Given the formal similarity between C-NOT and XOR operations, one might expect that we could use the configuration shown in Fig. 3.4 to produce $\psi \otimes \psi$ at the output (where ψ has the usual form $\psi = \alpha \psi_0 + \beta \psi_1$).

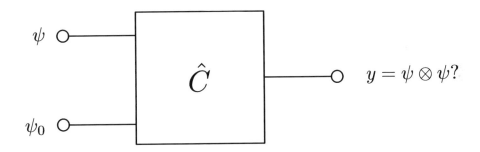

Figure 3.4: A hypothetical quantum "copy" operation using a C-NOT gate.

It turns out, however, that this is impossible, since

$$\hat{C}\left(\psi \otimes \psi_0\right) = \alpha\left(\psi_0 \otimes \psi_0\right) + \beta\left(\psi_1 \otimes \psi_1\right) = \alpha \Psi_{00} + \beta \Psi_{11} \tag{3.78}$$

is clearly different from

$$\psi \otimes \psi = (\alpha\psi_0 + \beta\psi_1) \otimes (\alpha\psi_0 + \beta\psi_1) = \alpha^2 \Psi_{00} + \alpha\beta \Psi_{01} + \alpha\beta \Psi_{10} + \beta^2 \Psi_{11} \tag{3.79}$$

except in the special case when $\alpha = 1$, $\beta = 0$, or $\alpha = 0$, $\beta = 1$. As a result, the C-NOT gate shown in Fig. 3.4 allows us to copy only states ψ_0 and ψ_1, but *not* a general state of the form $\psi = \alpha\psi_0 + \beta\psi_1$.

C-NOT *Gates and Information Entropy*

Although the C-NOT gate is reversible and allows us to recover the original state of the two qubits, it nevertheless "hides" a certain amount of information about this system. To see that a bit more clearly, let us once again examine the configuration shown in Fig. 3.1, where the control qubit $\psi_a = \alpha\psi_0 + \beta\psi_1$ is in a state of superposition while the target qubit is in state ψ_0. The initial state of the two particle system before the $\hat{C}$ operation is performed is

$$\Psi = \alpha\left(\psi_0 \otimes \psi_0\right) + \beta\left(\psi_1 \otimes \psi_0\right) = \alpha\Psi_{00} + \beta\Psi_{10} \tag{3.80}$$

which means that a measurement could show either 00 or 10. The amount of "hidden" information (or entropy) in this case is exactly 1 bit, since there are two possible outcomes, and we must ask one Yes/No question to establish which one will materialize. Note, however, that all of the uncertainty in the system is associated with the *control particle*, since the target particle is in a definite state (it always produces a 0 when a measurement is made).

What happens after the C-NOT operation is performed? Recalling that $\hat{C}(\Psi_{00}) = \Psi_{00}$ and $\hat{C}(\Psi_{10}) = \Psi_{11}$, it follows that the aggregate state takes the form $\Phi = \hat{C}(\Psi) = \alpha\Psi_{00} + \beta\Psi_{11}$. The information entropy of the system is still 1 bit (since we have two possible outcomes, 00 or 11), but the uncertainty is now *shared* by the two particles. In other words, in the course of the C-NOT operation, the control particle "infects" the target particle with uncertainty (*both* of them can now produce a 0 or a 1 when a measurement is made).

This "spreading of uncertainty" has a nice interpretation in terms of *mutual entropy*. Before the C-NOT operation takes place, the control particle has an uncertainty of 1 bit, which we can express as $H(X) = 1$. The target particle, on the other hand, is in a definite state, so its uncertainty is $H(Y) = 0$. According to equation (1.62), the mutual entropy associated with this system can be calculated as $I(X|Y) = H(X) - H(X|Y)$, where $H(X|Y)$ represents the conditional entropy (i.e., the amount of uncertainty about X *after Y is known*). Since the state of the control qubit is initially unaffected by our knowledge of the state of the target qubit, it follows that $H(X|Y) = H(X)$. In that case, we obtain $I(X|Y) = H(X) - H(X) = 0$, which indicates that Y conveys no information about X.

After the C-NOT operation, the composite wave function assumes the form $\Psi = \alpha\Psi_{00} + \beta\Psi_{11}$. The conditional entropy $H(X|Y)$ now becomes zero, since the two particles will have exactly the same state after a measurement is made

3.2. QUANTUM COMPUTING

(in other words, there is *no* uncertainty about X once we know Y). On the other hand, we know that measurements made on either of the two particles can produce a 1 or a 0, so their *individual* uncertainties are $H(X) = H(Y) = 1$. As a result, we have that $I(X|Y) = H(X) = 1$, which means that the mutual entropy has increased from zero to one bit.

3.2.3 Classical Computations on Quantum Computers

The fact that $\hat{X}$, $\hat{Z}$, $\hat{H}$ and $\hat{C}$ are unitary operators suggests that this is an inherent property of all quantum gates. To see why this property is important, we should first recognize that any operation performed on a system of n qubits is equivalent to a transformation of the form

$$\begin{bmatrix} \rho_1 \\ \vdots \\ \rho_{2^n} \end{bmatrix} = U \cdot \begin{bmatrix} \alpha_1 \\ \vdots \\ \alpha_{2^n} \end{bmatrix} \tag{3.81}$$

where U is a symmetric matrix of dimension $2^n \times 2^n$. As a result, we have that

$$\sum_{i=1}^{2^n} |\rho_i|^2 = \rho^T \rho = \alpha^T U^T U \alpha = \alpha^T U^2 \alpha \tag{3.82}$$

If operator $\hat{U}$ is unitary, the corresponding matrix U must satisfy $U^2 = I$, which ensures that the normalization condition

$$\sum_{i=1}^{2^n} |\rho_i|^2 = \sum_{i=1}^{2^n} |\alpha_i|^2 = 1 \tag{3.83}$$

continues to hold after the transformation. It also ensures that applying operator $\hat{U}$ twice will always return the system to its original state (which is why we say that quantum gates must be *reversible*).

Because most conventional logic gates (such as NAND and XOR gates, for example) are *irreversible*, it is impossible to directly map their operation onto a quantum computer. In order to do that, we would first have to transform the original digital circuit into an equivalent one that consists exclusively of *reversible gates*. It turns out that this requirement is met by the so-called *Toffoli gate*, whose schematic description is shown in Fig. 3.5.

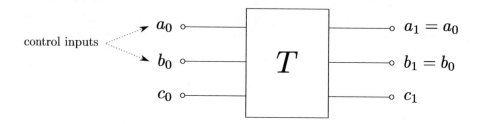

Figure 3.5: A Toffoli gate.

In this diagram, a_0 and b_0 represent control inputs that are *automatically replicated* at the output, while c_1 differs from c_0 only when $a_0 = b_0 = 1$. The truth table for this gate is shown in Table 3.3 (the highlighted bits correspond to the two scenarios in which $c_1 \neq c_0$).

a_0	b_0	c_0	a_1	b_1	c_1
0	0	0	0	0	0
0	0	1	0	0	1
0	1	0	0	1	0
0	1	1	0	1	1
1	0	0	1	0	0
1	0	1	1	0	1
1	1	**0**	1	1	**1**
1	1	**1**	1	1	**0**

Table 3.3. The truth table for a Toffoli gate.

To see why this operation is reversible, let us consider the cascade connection of two Toffoli gates shown in Fig. 3.6.

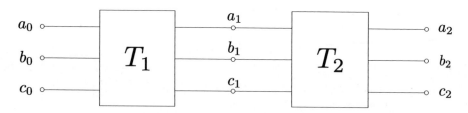

Figure 3.6: A configuration that performs two consecutive Toffoli operations.

The control inputs in the first stage are a_0 and b_0, while a_1 and b_1 assume this role in the second stage. The combined truth table shown below indicates that outputs a_2, b_2 and c_2 are identical to a_0, b_0 and c_0, which means

3.2. QUANTUM COMPUTING

that the original inputs can be recovered by performing two consecutive Toffoli operations.

a_0	b_0	c_0	a_1	b_1	c_1	a_2	b_2	c_2
0	0	0	0	0	0	0	0	0
0	0	1	0	0	1	0	0	1
0	1	0	0	1	0	0	1	0
0	1	1	0	1	1	0	1	1
1	0	0	1	0	0	1	0	0
1	0	1	1	0	1	1	0	1
1	1	0	1	1	**1**	1	1	**0**
1	1	1	1	1	**0**	1	1	**1**

Table 3.4. Combined truth table for a pair of Toffoli gates.

In order to show that *any* logic circuit can be realized using a combination of Toffoli gates, it suffices to demonstrate that this gate can simulate the NAND operation, since this type of gate is known to be *universal* (i.e., it represents the basic building block from which all other standard gates can be built). The diagram in Fig. 3.7 shows how this can be done.

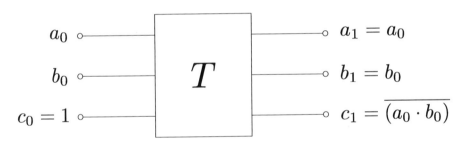

Figure 3.7: A NAND operation using a Toffoli gate.

The possible input and output combinations for this circuit are provided in Table 3.5, which indicates that the behavior of output c_1 conforms to the NAND function $c_1 = \overline{a_0 \cdot b_0}$. Note that this table has only *half* of the possible input combinations of a regular Toffoli gate, since input c_0 is fixed at 1.

a_0	b_0	c_0	a_1	b_1	c_1
0	0	1	0	0	1
0	1	1	0	1	1
1	0	1	1	0	1
1	1	1	1	1	**0**

Table 3.5. Truth table for the configuration in Fig. 3.7.

What would a *quantum* Toffoli gate look like? A schematic diagram of such a gate is shown in Fig. 3.8, where qubits ψ_a, ψ_b and ψ_c are assumed to share a common basis $\{\psi_0, \psi_1\}$.

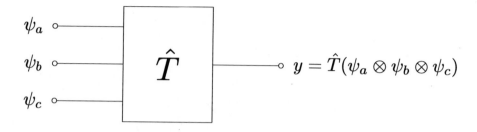

Figure 3.8: A quantum Toffoli gate.

When operator $\hat{T}$ is applied to the composite function

$$\Psi = \psi_a \otimes \psi_b \otimes \psi_c = \gamma_0 \Psi_{000} + \gamma_1 \Psi_{001} + \gamma_2 \Psi_{010} + \gamma_3 \Psi_{011} + \\ + \gamma_4 \Psi_{100} + \gamma_5 \Psi_{101} + \gamma_6 \Psi_{110} + \gamma_7 \Psi_{111} \tag{3.84}$$

we obtain

$$\Phi = \hat{T}(\Psi) = \gamma_0 \Psi_{000} + \gamma_1 \Psi_{001} + \gamma_2 \Psi_{010} + \gamma_3 \Psi_{011} + \\ + \gamma_4 \Psi_{100} + \gamma_5 \Psi_{101} + \gamma_7 \Psi_{110} + \gamma_6 \Psi_{111} \tag{3.85}$$

Note that the only components that change in this process are the ones where the first two indices are both equal to 1. For those components, the third index will "flip", producing $\hat{T}(\Psi_{110}) = \Psi_{111}$ and $\hat{T}(\Psi_{111}) = \Psi_{110}$, respectively, while all other coefficients remain unaffected. This means that coefficients γ_6 and γ_7 effectively "trade places.

If we denote the coefficients associated with function Φ by $\{\rho_1, \rho_2, \ldots, \rho_7\}$, the relationship between $\{\rho_i\}$ and $\{\gamma_i\}$ can be expressed as

$$\begin{bmatrix} \rho_0 \\ \rho_1 \\ \rho_2 \\ \rho_3 \\ \rho_4 \\ \rho_5 \\ \rho_6 \\ \rho_7 \end{bmatrix} = \begin{bmatrix} 1 & 0 & 0 & 0 & 0 & 0 & 0 & 0 \\ 0 & 1 & 0 & 0 & 0 & 0 & 0 & 0 \\ 0 & 0 & 1 & 0 & 0 & 0 & 0 & 0 \\ 0 & 0 & 0 & 1 & 0 & 0 & 0 & 0 \\ 0 & 0 & 0 & 0 & 1 & 0 & 0 & 0 \\ 0 & 0 & 0 & 0 & 0 & 1 & 0 & 0 \\ 0 & 0 & 0 & 0 & 0 & 0 & 0 & 1 \\ 0 & 0 & 0 & 0 & 0 & 0 & 1 & 0 \end{bmatrix} \begin{bmatrix} \gamma_0 \\ \gamma_1 \\ \gamma_2 \\ \gamma_3 \\ \gamma_4 \\ \gamma_5 \\ \gamma_6 \\ \gamma_7 \end{bmatrix} \tag{3.86}$$

It is not difficult to verify that the matrix T in (3.86) satisfies $T^2 = I$, which

3.2. QUANTUM COMPUTING

means that $\hat{T}$ is a *unitary* operator.

The connection between classical and quantum Toffoli gates becomes apparent if we represent the indices of coefficients $\rho_0 \div \rho_7$ and $\gamma_0 \div \gamma_7$ in *binary form* as $(a_0\, b_0\, c_0)$ and $(a_1\, b_1\, c_1)$, respectively. Observing that ρ_6 maps into γ_7 and ρ_7 into γ_6, we obtain the following truth table, which is identical to the one in Table 3.3.

	a_0	b_0	c_0			a_1	b_1	c_1
$\rho_0:$	0	0	0	$\rightarrow$	$\gamma_0:$	0	0	0
$\rho_1:$	0	0	1	$\rightarrow$	$\gamma_1:$	0	0	1
$\rho_2:$	0	1	0	$\rightarrow$	$\gamma_2:$	0	1	0
$\rho_3:$	0	1	1	$\rightarrow$	$\gamma_3:$	0	1	1
$\rho_4:$	1	0	0	$\rightarrow$	$\gamma_4:$	1	0	0
$\rho_5:$	1	0	1	$\rightarrow$	$\gamma_5:$	1	0	1
$\rho_6:$	1	1	0	$\rightarrow$	$\gamma_7:$	1	1	1
$\rho_7:$	1	1	1	$\rightarrow$	$\gamma_6:$	1	1	0

Table 3.6. The correspondence between classical and quantum Toffoli gates.

3.2.4 Quantum Algorithms

Quantum algorithms can perform certain computational tasks far more efficiently than conventional computer programs. To see how this can be done, let us consider the following simple problem, in which $f(x)$ is an unknown function that can take only values 0 or 1. We will assume that variable x is also limited to values 0 and 1, and our objective will be to establish whether or not $f(0) = f(1)$ (in other words, we would like to know if function f is *constant*).

If we were to approach this problem in the conventional way, we would have to evaluate function $f(x)$ *twice* (for $x = 0$ and for $x = 1$). It turns out, however, that quantum computers can do this with a *single* operation. In the following, we will describe an algorithm that can accomplish this.

Deutsch's Algorithm

Let us define an operator $\hat{U}_f$ that transforms the composite function $\psi_x \otimes \psi_y$ in the following way

$$\hat{U}_f(\psi_x \otimes \psi_y) \rightarrow \psi_x \otimes [f(x) \oplus \psi_y] \qquad (3.87)$$

where

$$f(x) \oplus \psi_y = \begin{cases} \psi_y, & \text{if } f(x) = 0 \\ \psi_{\bar{y}}, & \text{if } f(x) = 1 \end{cases} \qquad (3.88)$$

(the index $\bar{y}$ in (3.88) represents the complement of y, so $\bar{y} = 1$ when $y = 0$ and vice versa). If we choose ψ_x and ψ_y as

$$\psi_x = \psi_+ = \frac{1}{\sqrt{2}}(\psi_0 + \psi_1) \tag{3.89}$$

and

$$\psi_y = \psi_- = \frac{1}{\sqrt{2}}(\psi_0 - \psi_1) \tag{3.90}$$

we have that

$$\psi_x \otimes \psi_y = \frac{1}{2}(\psi_0 + \psi_1) \otimes (\psi_0 - \psi_1) = \frac{1}{2}(\psi_0 \otimes \psi_0) - \frac{1}{2}(\psi_0 \otimes \psi_1) + \\ + \frac{1}{2}(\psi_1 \otimes \psi_0) - \frac{1}{2}(\psi_1 \otimes \psi_1) \tag{3.91}$$

Applying operator $\hat{U}_f$ to (3.91) will then produce

$$\hat{U}_f(\psi_x \otimes \psi_y) = \frac{1}{2}\hat{U}_f(\psi_0 \otimes \psi_0) - \frac{1}{2}\hat{U}_f(\psi_0 \otimes \psi_1) + \\ + \frac{1}{2}\hat{U}_f(\psi_1 \otimes \psi_0) - \frac{1}{2}\hat{U}_f(\psi_1 \otimes \psi_1) \tag{3.92}$$

The definition provided in (3.87) implies that the four terms in expression (3.92) have the form:

$$\begin{aligned} \hat{U}_f(\psi_0 \otimes \psi_0) &= \psi_0 \otimes [f(0) \oplus \psi_0] \\ \hat{U}_f(\psi_0 \otimes \psi_1) &= \psi_0 \otimes [f(0) \oplus \psi_1] \\ \hat{U}_f(\psi_1 \otimes \psi_0) &= \psi_1 \otimes [f(1) \oplus \psi_0] \\ \hat{U}_f(\psi_1 \otimes \psi_1) &= \psi_1 \otimes [f(1) \oplus \psi_1] \end{aligned} \tag{3.93}$$

Substituting this into (3.92), we have

$$\hat{U}_f(\psi_x \otimes \psi_y) = \frac{1}{2}\psi_0 \otimes [f(0) \oplus \psi_0] - \frac{1}{2}\psi_0 \otimes [f(0) \oplus \psi_1] + \\ + \frac{1}{2}\psi_1 \otimes [f(1) \oplus \psi_0] - \frac{1}{2}\psi_1 \otimes [f(1) \oplus \psi_1] \tag{3.94}$$

If we group the terms, (3.94) can be rewritten as

$$\hat{U}_f(\psi_x \otimes \psi_y) = \frac{1}{2}[\psi_0 \otimes (f(0) \oplus \psi_0 - f(0) \oplus \psi_1) + \\ + \psi_1 \otimes (f(1) \oplus \psi_0 - f(1) \oplus \psi_1)] \tag{3.95}$$

3.2. QUANTUM COMPUTING

This is convenient because terms of the form $f(x) \oplus \psi_0 - f(x) \oplus \psi_1$ can be expressed as

$$f(x) \oplus \psi_0 - f(x) \oplus \psi_1 = (-1)^{f(x)}(\psi_0 - \psi_1) \quad (3.96)$$

for any value of $f(x)$. To see why this is so, we should first recall that the operation $f(x) \oplus \psi_y$ leaves ψ_y unchanged when $f(x) = 0$. As a result, for $f(x) = 0$ we have

$$f(x) \oplus \psi_0 - f(x) \oplus \psi_1 = \psi_0 - \psi_1 = (-1)^0(\psi_0 - \psi_1) \quad (3.97)$$

On the other hand, if $f(x) = 1$ we know that $f(x) \oplus \psi_y = \psi_{\bar{y}}$, so in that case we have

$$f(x) \oplus \psi_0 - f(x) \oplus \psi_1 = \psi_1 - \psi_0 = -(\psi_0 - \psi_1) = (-1)^1(\psi_0 - \psi_1) \quad (3.98)$$

It is not difficult to see that both (3.97) and (3.98) conform to the general expression (3.96), which allows us to rewrite (3.95) as

$$\hat{U}_f(\psi_x \otimes \psi_y) = \frac{1}{2}\left[(-1)^{f(0)}\psi_0 \otimes (\psi_0 - \psi_1) + (-1)^{f(1)}\psi_1 \otimes (\psi_0 - \psi_1)\right] \quad (3.99)$$

For the purposes of this derivation, it will be convenient to factor out the term $(-1)^{f(0)}$ in (3.99), which produces

$$\hat{U}_f(\psi_x \otimes \psi_y) = \frac{1}{2}(-1)^{f(0)}[\psi_0 \otimes (\psi_0 - \psi_1) + \\ +(-1)^{f(1)-f(0)}\psi_1 \otimes (\psi_0 - \psi_1)] \quad (3.100)$$

The reason why this is helpful becomes clear if we examine Table 3.7, in which functions $g = f(1) - f(0)$ and $h = f(0) \oplus f(1)$ are evaluated for different values of $f(0)$ and $f(1)$ (note that $\oplus$ denotes the logic XOR operation in this context).

$f(0)$	$f(1)$	$f(0) \oplus f(1)$	$f(1) - f(0)$
0	0	0	0
0	1	1	1
1	0	1	-1
1	1	0	0

Table 3.7. Operations with $f(0)$ and $f(1)$.

Using Table 3.7, it is easily verified that

$$(-1)^{f(1)-f(0)} = \begin{cases} 1, & \text{if } f(0) = f(1) \\ -1, & \text{if } f(0) \neq f(1) \end{cases} \quad (3.101)$$

and
$$(-1)^{f(0)\oplus f(1)} = \begin{cases} 1, & \text{if } f(0) = f(1) \\ -1, & \text{if } f(0) \neq f(1) \end{cases} \quad (3.102)$$

which allows us to replace $(-1)^{f(1)-f(0)}$ with $(-1)^{f(1)\oplus f(0)}$. Substituting this into (3.100) produces

$$\hat{U}_f(\psi_x \otimes \psi_y) = \frac{1}{2}(-1)^{f(0)}[\psi_0 \otimes (\psi_0 - \psi_1) + \\ +(-1)^{f(0)\oplus f(1)}\psi_1 \otimes (\psi_0 - \psi_1)] \quad (3.103)$$

which becomes

$$\hat{U}_f(\psi_x \otimes \psi_y) = \frac{1}{2}(-1)^{f(0)}\left[(\psi_0 + (-1)^{f(0)\oplus f(1)}\psi_1) \otimes (\psi_0 - \psi_1)\right] \quad (3.104)$$

after grouping the terms.

The practical significance of this result becomes apparent if we rewrite (3.104) as

$$\hat{U}_f(\psi_x \otimes \psi_y) = \psi_A \otimes \psi_B \quad (3.105)$$

where

$$\psi_A = \alpha\psi_0 + \beta\psi_1 \quad (3.106)$$

and

$$\psi_B = \frac{1}{\sqrt{2}}(-1)^{f(0)}(\psi_0 - \psi_1) \quad (3.107)$$

with

$$\alpha = \frac{1}{\sqrt{2}} \quad \text{and} \quad \beta = \frac{(-1)^{f(0)\oplus f(1)}}{\sqrt{2}} \quad (3.108)$$

If we now pass function ψ_A through a Hadamard gate, we get

$$\psi_{\text{out}} = \frac{\alpha}{\sqrt{2}}(\psi_0 + \psi_1) + \frac{\beta}{\sqrt{2}}(\psi_0 - \psi_1) \quad (3.109)$$

When function $f(x)$ is *constant* (i.e., when $f(0) = f(1)$), Table 3.7 indicates that $f(0) \oplus f(1) = 0$. Consequently,

$$\beta = \frac{(-1)^{f(0)\oplus f(1)}}{\sqrt{2}} = \frac{1}{\sqrt{2}} \quad (3.110)$$

and

$$\psi_{\text{out}} = \frac{1}{2}(\psi_0 + \psi_1) + \frac{1}{2}(\psi_0 - \psi_1) = \psi_0 \quad (3.111)$$

When $f(0) \neq f(1)$, we have that $f(0) \oplus f(1) = 1$, so β becomes

$$\beta = \frac{(-1)^{f(0)\oplus f(1)}}{\sqrt{2}} = -\frac{1}{\sqrt{2}} \quad (3.112)$$

3.2. QUANTUM COMPUTING

and we have
$$\psi_{\text{out}} = \frac{1}{2}(\psi_0 + \psi_1) - \frac{1}{2}(\psi_0 - \psi_1) = \psi_1 \tag{3.113}$$

Equations (3.111) and (3.113) tell us that we can determine whether or not $f(x)$ is a constant with a *single* measurement of ψ_{out}. As noted earlier, any classical algorithm would require *two* separate evaluations of function $f(x)$, as well as a comparison of the obtained results. The power of quantum parallelism becomes even more evident if Deutsch's algorithm is generalized to include the case when x can take any value in the set $N = \{0, 1, 2, \ldots, 2^{n-1}\}$. Function $f(x)$ is once again allowed to take only values 0 and 1, but we additionally assume that it can either be *constant* for all possible x, or can produce an *equal number* of zeros and ones on set N (such a function is said to be *balanced*). The so-called Deutsch-Josza algorithm can determine whether $f(x)$ is constant or balanced using a *single* measurement, while a conventional algorithm would require at least $2^{n-1} + 1$ independent tests to do this. Because the number of such evaluations grows exponentially with n, we can safely say that such a task would eventually overwhelm any digital computer, no matter how powerful it is.

Chapter 4

Notes and References

4.1 Notes for Chapters 1-3

1. For a proof of this result, see, e.g. William Feller, *An Introduction to Probability Theory and Its Applications*, Vol. II, WSE, 2008.

2. For a proof of this theorem, see e.g.: Thomas Cover and Joy Thomas, *Elements of Information Theory*, Wiley-Interscience, 2006. Shannon's original result was published in: C. Shannon, "Communication in the presence of noise", *Proceedings of the Institute of Radio Engineers*, **37**, pp. 10 – 21, 1949.

3. See, e.g. Venkatarama Krishnan and Kavitha Chandra, *Probability and Random Processes*, Wiley, 2015.

4. It is important to recognize that the process of synthesizing proteins has a *feedback loop*, since the ribosome (where the translation of mRNA takes place) is itself made up of proteins. Proteins also act as catalysts for the chemical reactions that produce nucleic acids, which raises the question of how such complex compounds could have emerged in the "prebiotic" stage of evolution. A plausible explanation is that primitive nucleic acids may have had the ability to self-reproduce even in the absence of proteins. A number of models suggest that something like this might be possible in systems that are capable of *autocatalysis* (see Section 8.1.2).

5. For a thorough discussion on this subject, see, e.g.: Gregory Chaitin, *Algorithmic Information Theory*, Cambridge University Press, 1987.

6. Although mathematical models tend to treat RNA and DNA sequences as "random" strings of symbols, we must keep in mind that they are produced in nature, and that this happens over a finite period of time. We therefore cannot rule out the possibility that there is some underlying "program" that generates them, whose structure we may never know. Whether or not this "program" is as long as the sequences that it produces is impossible to tell, so the assumption that genetic sequences are random is clearly not something that we can formally prove.

7. Hubert Yockey, *Information Theory, Evolution and the Origins of Life*, Cambridge University Press, 2011.

8. A more accurate model must also take into account the fact that proteins sometimes retain their functionality even though a "wrong" amino acid appears in a particular position in the chain (proteins are actually quite resilient with respect to amino acid substitutions). In order to account for this possibility, we would have to set $\alpha = 0$ for any such occurrence, which considerably complicates the calculation. This added level of complexity is necessary, however, if we want to precisely determine how much information is lost as a result of mutations.

9. See. e.g. Seth Lloyd, *Programming the Universe*, Knopf, 2006.

10. It has been shown that Shor's algorithm for prime factorization runs in polynomial time on a quantum computer. A description of this algorithm can be found in: P. Shor, "Polynomial-time algorithms for prime factorization and discrete logarithms on a quantum computer", *SIAM Journal on Computing*, **26**, pp. 1484-1509, 1997. For a discussion of possible physical implementations, see: C-Y. Lu, D. Browne, T., Yang and J-W. Pan, "Demonstration of a compiled version of Shor's quantum factoring algorithm using photonic qubits", *Physical Review Letters*, **99**, 250504, 2007.

11. N. Margolus and L. Levitin, "The maximum speed of dynamical evolution", *Physica D*, **120**, pp. 188–195, 1998.

12. It is interesting to note in this context that quantum computers *always* transform their qubits at the maximal possible rate (unlike ordinary computers, for which this is only a theoretical limit).

4.2 Further Reading

Classical Information Theory

1. A. I. Khinchin, *Mathematical Foundations of Information Theory*, Dover Publication, 1957.

2. Claude Shannon and Warren Weaver, *The Mathematical Theory of Communication*, The University of Illinois Press, 1971.

3. John Pierce, *An Introduction to Information Theory*, Dover Publications, 1980.

4. Robert Ash, *Information Theory*, Dover Publications, 1990.

5. Thomas Cover and Joy Thomas, *Elements of Information Theory*, Wiley-Interscience, 2006.

Information in Biology

1. Bernd-Olaf Küppers, *Information and the Origins of Life*, MIT Press, 1990.

2. John Maynard-Smith and Eors Szathmary, *The Major Transitions in Evolution*, Oxford University Press, 1998.

3. Neil Jones and Pavel Pevzner, *An Introduction to Bioinformatics Algorithms*, MIT Press, 2004.

4. Hubert Yockey, *Information Theory, Evolution and the Origins of Life*, Cambridge University Press, 2011.

5. Samuli Niiranen and Andre Ribeiro (Eds.), *Information Processing and Biological Systems*, Springer , 2011.

6. John Avery, *Information Theory and Evolution*, World Scientific, 2012.

7. Robert Marks, John Sanford and Michael Behe (Eds.), *Biological Information: New Perspectives*, World Scientific, 2013.

Quantum Information and Computing

1. Michael Nielsen and Isaac Chuang, *Quantum Computation and Quantum Information*, Cambridge University Press, 2000.

2. Arthur Pittenger, *An Introduction to Quantum Computing Algorithms*, Birkhauser, 2001.

3. Philip Kaye, Raymonde Laflamme and Michele Mosca, *An Introduction to Quantum Computing*, Oxford University Press, 2007.

4. Mark Wilde, *Quantum Information Theory*, Cambridge University Press, 2013.

5. Christopher Timpson, *Quantum Information Theory and the Foundations of Quantum Mechanics*, Oxford University Press, 2013.

Part II
Self-Similarity and Fractals

Chapter 5

Self-Similarity

5.1 Power Laws and Spectral Density

In planar geometry, two objects are said to be *similar* if one of them can be obtained from the other by successively applying a combination of four basic "shape preserving" transformations: *scaling, rotation, translation* and *reflection*. The notion of *self-similarity* is related to this idea, but is not as easy to define. This property is often associated with objects whose parts can produce an exact copy of the whole when appropriately enlarged. The difficulty with such a description (which is simple and intuitive) is that it is not sufficiently precise, since phrases like "parts" and "appropriately enlarged" fail to specify *which* segment (or segments) should be chosen, or what the scaling factor should be.

A possible way to eliminate ambiguities of this sort is based on the observation that complex objects (and complex physical processes) can be characterized by describing how some relevant property Q changes as it is measured with an increasingly finer resolution s. In the special case when the relationship between Q and s takes the form of a *power law*

$$Q(s) = Ks^b \tag{5.1}$$

it is appropriate to say that the object is self-similar, because its essential features are invariant with respect to scaling.

To see why this is so, let us consider what happens when we increase (or decrease) the resolution by a factor of a. In mathematical terms, this operation is equivalent to replacing s with as, which produces

$$Q(as) = K(as)^b = Ka^b s^b = a^b Q(s) \tag{5.2}$$

What expression (5.2) tells us is that $Q(as)$ is just a scaled version of the original function, which differs from it by a constant multiplier a^b. As such, it retains all the relevant characteristics of $Q(s)$.

It is worth mentioning in this context that power laws are not the only type of function that exhibits this sort of invariance with respect to scaling. Something similar also holds for relationships of the form

$$Q(s) = K s^b f\left(\frac{\log s}{\log w}\right) \tag{5.3}$$

where K and w are given constants, and f is a periodic function that satisfies

$$f(x) = f(1+x) \tag{5.4}$$

To verify that this is indeed the case, it suffices to observe that

$$Q(as) = K a^b s^b f\left(\frac{\log as}{\log w}\right) = K a^b s^b f\left(\frac{\log a + \log s}{\log w}\right) \tag{5.5}$$

In the special case when $a = w$, (5.5) reduces to

$$Q(ws) = K w^b s^b f\left(1 + \frac{\log s}{\log w}\right) = K w^b s^b f\left(\frac{\log s}{\log w}\right) \tag{5.6}$$

which is equivalent to

$$Q(ws) = w^b Q(s) \tag{5.7}$$

We should keep in mind, however, that this condition is far more restrictive than (5.1), since $Q(as)$ is proportional to $Q(s)$ *only* when $a = w$ (which is a very special choice of scaling). Power laws, on the other hand, have no such constraints, and satisfy the self-similarity condition for *any* a.

Power laws have particularly interesting consequences in cases when they describe how the length of an object is related to the resolution of the "ruler" that is used to measure it. For ordinary geometric figures such as lines and smooth curves, using increasingly finer rulers makes little difference in the overall result, since their length $Q(s)$ converges to some number

$$\bar{Q} = \lim_{s \to 0} Q(s) \tag{5.8}$$

when $s \to 0$. This number represents the "correct" value of Q, which we can approximate to a desired precision by using a sufficiently accurate measurement device. However, if we assume that $Q(s)$ satisfies equation (5.1) with a *negative* exponent b, this quantity will become infinitely large when s approaches zero. Under such circumstances, it makes no sense to speak of a "correct" value for Q - what we get will depend on the precision with which we measure, and *all* answers are equally meaningful. We will see examples of this unusual property in Chapter 6, which features a number of fractals whose length conforms to a power law, and grows unboundedly as the resolution of the "ruler" becomes finer.

5.1. POWER LAWS AND SPECTRAL DENSITY

The notion of self-similarity becomes considerably more complicated when we examine forms that appear in nature. Here we must consider the possibility that scaling could produce only distorted copies of the original, which share some of its general features but have very different details. In such cases it is more appropriate to speak of *statistical self-similarity*, in the sense that the original and the scaled "copies" have the same mean, variance, moments, etc.

To get a better sense for what this concept means, in the following we will briefly examine how it applies to time-varying signals. We will first consider deterministic processes (since these are easier to deal with) and will then extend the ideas that we develop to random signals whose frequency domain behavior takes the form of a power law.

Deterministic Processes

Two quantities that are commonly used to describe a deterministic signal $x(t)$ are its *autocorrelation function* and its *power spectral density*. The autocorrelation function represents a measure for how closely $x(t)$ and $x(t - \tau)$ are related, and is defined as

$$R_x(\tau) = \lim_{T \to \infty} \frac{1}{T} \int_{-T/2}^{T/2} x(t)x(t - \tau)dt \qquad (5.9)$$

An important characteristic of this function is its symmetry with respect to time reversals, which can be formally expressed as $R_x(\tau) = R_x(-\tau)$. We can easily verify this property by introducing a new variable $\sigma = t - \tau$ into (5.9), which produces

$$R_x(\tau) = \lim_{T \to \infty} \frac{1}{T} \int_{-T/2-\tau}^{T/2-\tau} x(\sigma + \tau)x(\sigma)d\sigma \qquad (5.10)$$

Since $T/2 >> \tau$ when $T \to \infty$, this integral becomes

$$R_x(\tau) = \lim_{T \to \infty} \frac{1}{T} \int_{-T/2}^{T/2} x(\sigma)x(\sigma + \tau)d\sigma = R_x(-\tau) \qquad (5.11)$$

The Fourier transform of the autocorrelation function

$$S(f) = \int_{-\infty}^{\infty} R_x(\tau)e^{-j2\pi f\tau}d\tau \qquad (5.12)$$

is referred to as the *power spectral density* of signal $x(t)$. The following lemma shows that this quantity can be equivalently expressed as

$$S(f) = \lim_{T \to \infty} \frac{1}{T} |X_T(f)|^2 \qquad (5.13)$$

where $X_T(f)$ represents the Fourier transform of function

$$x_T(t) = \begin{cases} x(t), & -T/2 \le t \le T/2 \\ 0, & \text{otherwise} \end{cases} \qquad (5.14)$$

Lemma 5.1. Let $x_T(t)$ denote the function defined in (5.14) (which is commonly referred to as the *truncation* of $x(t)$ on interval $[-T/2 \ T/2]$). Then,

$$S(f) = \lim_{T \to \infty} \frac{1}{T} |X_T(f)|^2 \qquad (5.15)$$

Proof. The proof of this lemma hinges on two fundamental properties of Fourier transforms:

Property 1. The Fourier transforms of functions $g(t)$ and $g(-t)$ are related as

$$F\{g(-t)\} = F^*\{g(t)\} \qquad (5.16)$$

Property 2. Let

$$y(t) = \int_{-\infty}^{\infty} g(t - \alpha) h(\alpha) d\alpha \qquad (5.17)$$

denote the convolution of functions g and h. Then, the Fourier transform of $y(t)$ satisfies

$$F\{y(t)\} = F\{g(t)\} \cdot F\{h(t)\} \qquad (5.18)$$

According to (5.17), the convolution of $g(t) = x_T(t)$ and $h(t) = x_T(-t)$ takes the form

$$y(t) = \int_{-\infty}^{\infty} x_T(t - \alpha) x_T(-\alpha) d\alpha = \int_{-\infty}^{\infty} x_T(t + \sigma) x_T(\sigma) d\sigma \qquad (5.19)$$

5.1. POWER LAWS AND SPECTRAL DENSITY

Since $x_T(\sigma) = x(\sigma)$ for $-T/2 \leq \sigma \leq T/2$ and is zero everywhere else, (5.19) can be simplified as

$$y(t) = \int_{-T/2}^{T/2} x_T(t+\sigma) x_T(\sigma) d\sigma \tag{5.20}$$

The Fourier transform of $y(t)$ then becomes

$$F\{y(t)\} = \int_{-\infty}^{\infty} y(\tau) e^{-j2\pi f \tau} d\tau =$$

$$= \int_{-\infty}^{\infty} \left[\int_{-T/2}^{T/2} x_T(\tau+\sigma) x_T(\sigma) d\sigma \right] e^{-j2\pi f \tau} d\tau \tag{5.21}$$

In order to connect this expression to the power spectral density, we should observe that

$$|X_T(f)|^2 = X_T(f) \cdot X_T^*(f) = F\{x_T(t)\} \cdot F\{x_T(-t)\} \tag{5.22}$$

by virtue of Property 1. Since $y(t)$ represents the convolution of $x_T(t)$ and $x_T(-t)$, we also know that

$$F\{x_T(t)\} \cdot F\{x_T(-t)\} = F\{y(t)\} \tag{5.23}$$

This allows us to combine (5.21), (5.22) and (5.23), and express $|X_T(f)|^2$ as

$$|X_T(f)|^2 = F\{y(t)\} = \int_{-\infty}^{\infty} \left[\int_{-T/2}^{T/2} x_T(\tau+\sigma) x_T(\sigma) d\sigma \right] e^{-j2\pi f \tau} d\tau \tag{5.24}$$

Dividing both sides of (5.24) by T and taking the limit when $T \to \infty$, we obtain

$$\lim_{T \to \infty} \frac{1}{T} |X_T(f)|^2 = \int_{-\infty}^{\infty} e^{-j2\pi f \tau} d\tau \cdot \lim_{T \to \infty} \frac{1}{T} \int_{-T/2}^{T/2} x_T(\sigma) x_T(\sigma+\tau) d\sigma \tag{5.25}$$

Observing that $x_T(\sigma) = x(\sigma)$ and $x_T(\sigma+\tau) = x(\sigma+\tau)$ when $T \to \infty$, the second integral in equation (5.25) becomes

$$\lim_{T \to \infty} \frac{1}{T} \int_{-T/2}^{T/2} x(\sigma) x(\sigma+\tau) d\sigma = R_x(-\tau) = R_x(\tau) \tag{5.26}$$

and the overall expression can be rewritten as

$$\lim_{T\to\infty}\frac{1}{T}|X_T(f)|^2 = \int_{-\infty}^{\infty} R_x(\tau)e^{-j2\pi f\tau}d\tau = S(f) \qquad (5.27)$$

which is what we set out to prove. **Q.E.D.**

Stochastic Processes

In order to extend these ideas to random processes, we should first recognize that the outcomes in this case are not numbers, but rather *deterministic functions*. These functions can be represented as

$$x(t,\rho) \qquad (5.28)$$

where ρ is a random variable that determines which particular one will be observed. In the following, we will assume that ρ takes a set of discrete values $\{\rho_i\}$ which conform to a known probability distribution. In that case, at any given point in time we can think of $x(t)$ as a random variable whose possible values are

$$\begin{aligned} x_1 &= x(t,\rho_1) \\ x_2 &= x(t,\rho_2) \\ &\vdots \\ x_n &= x(t,\rho_n) \end{aligned} \qquad (5.29)$$

(this situation is schematically illustrated in Fig. 5.1). In interpreting this figure, we should note that at some different time $t' \neq t$ function $x(t')$ has an entirely different set of possible values

$$\begin{aligned} x'_1 &= x(t',\rho_1) \\ x'_2 &= x(t',\rho_2) \\ &\vdots \\ x'_n &= x(t',\rho_n) \end{aligned} \qquad (5.30)$$

Since $x(t)$ is a random variable at each point in time, it makes sense to think of its expected value as a function of time. We can define this function as

$$E[x(t)] = x(t,\rho_1)P(\rho_1) + x(t,\rho_2)P(\rho_2) + \ldots + x(t,\rho_n)P(\rho_n) \qquad (5.31)$$

5.1. POWER LAWS AND SPECTRAL DENSITY

where $P(\rho_i)$ represents the probability that ρ will take value ρ_i. Given that each function $x(t, \rho_i)$ has its own Fourier transform

$$X(f, \rho_i) = \int_{-\infty}^{\infty} x(t, \rho_i) e^{-j2\pi ft} dt \qquad (5.32)$$

we can think of $X(f, \rho_i)$ as a random variable in its own right, whose expected value is given by

$$E[X(f)] = X(f, \rho_1)P(\rho_1) + X(f, \rho_2)P(\rho_2) + \ldots + X(f, \rho_n)P(\rho_n) \qquad (5.33)$$

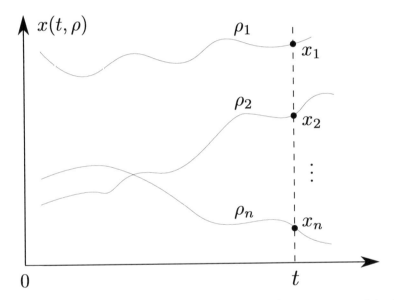

Figure 5.1: An ensemble of deterministic functions that correspond to different values of ρ_i.

Following the approach that we developed in the deterministic case, we can now define the power spectral density of a random signal $x(t)$ as

$$S(f) = \lim_{T \to \infty} \frac{1}{T} E\left[|X_T(f)|^2\right] \qquad (5.34)$$

where $X_T(f, \rho_i)$ corresponds to the truncated function

$$x_T(t, \rho_i) = \begin{cases} x(t, \rho_i), & -T/2 \leq t \leq T/2 \\ 0, & \text{otherwise} \end{cases} \qquad (5.35)$$

Recalling (5.25), this expression can be rewritten as

$$S(f) = \int_{-\infty}^{\infty} e^{-j2\pi f\tau} d\tau \cdot \lim_{T\to\infty} \frac{1}{T} \int_{-T/2}^{T/2} E\left[x(\sigma)x(\sigma+\tau)\right] d\sigma \qquad (5.36)$$

Since the term

$$R_x(\tau) = \lim_{T\to\infty} \frac{1}{T} \int_{-T/2}^{T/2} E\left[x(\sigma)x(\sigma+\tau)\right] d\sigma \qquad (5.37)$$

corresponds to the expected value of the autocorrelation function, (5.37) becomes

$$S(f) = \int_{-\infty}^{\infty} R_x(\tau) e^{-j2\pi f\tau} d\tau \qquad (5.38)$$

which is consistent with the way this quantity was defined for deterministic signals.

5.2 Self-Similarity in the Time Domain

Many types of random signals have power spectra of the form

$$S(f) = f^{-\beta} \qquad (5.39)$$

(at least, over a certain range of frequencies). The three most common ones are *white noise* ($\beta = 0$), *pink noise* ($\beta = 1$) and *brown noise* ($\beta = 2$) whose time-domain behavior is shown in Figs. 5.2 - 5.4.

The difference between these three types of signals can be seen by comparing their autocorrelation functions, which represent a measure of how closely $x(t)$ and $x(t+\tau)$ are related. In the case of white noise, $R_x(\tau)$ is zero for all $\tau > 0$, which means that there is no correlation whatsoever between the values of $x(t)$ at different points in time. Brown noise, on the other hand, is characterized by a significant correlation between $x(t)$ and $x(t+\tau)$ for small τ, which tends to disappear rather quickly as τ increases. In view of that, we can say that such signals have "short term memory", but no "long term memory".

When it comes to "memory", pink noise is in a category of its own, so to speak, because it allows for *both* short *and* long-term correlations between $x(t)$ and $x(t+\tau)$. As a result, such signals have interesting features on all time scales, and cannot be characterized by a "threshold" observation time below which we see nothing new.

5.2. SELF-SIMILARITY IN THE TIME DOMAIN

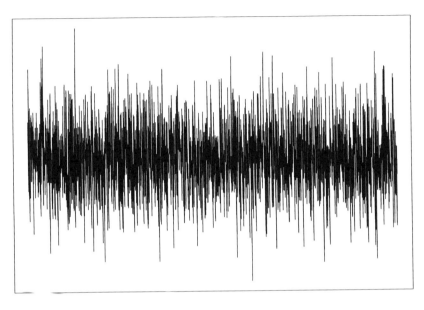

Figure 5.2: An example of white noise.

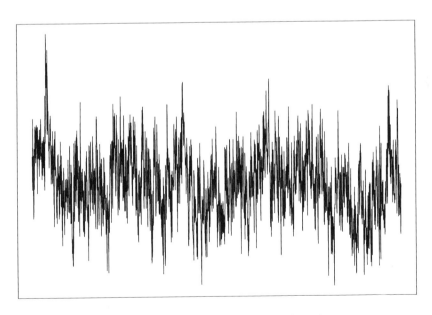

Figure 5.3: An example of pink noise.

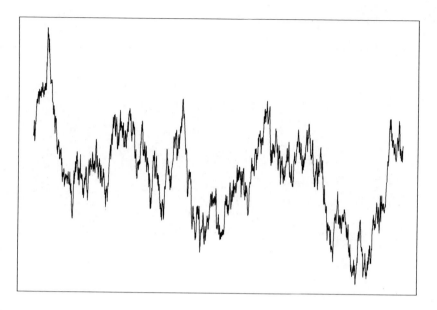

Figure 5.4: An example of brown noise.

To get a sense for why this is so, it is helpful to note that pink noise often arises in systems that allow for the simultaneous existence of multiple relaxation processes. A typical example of this sort would be a collection of particles whose energy levels have been temporarily elevated. The lifetime of such an "excited" state generally satisfies an exponential distribution, and the corresponding autocorrelation function has the form

$$R_x(\tau, \tau_C) = Ae^{-\tau/\tau_C}, \qquad \forall \tau > 0 \tag{5.40}$$

where the characteristic relaxation time τ_C can vary from one particle to another. The power spectrum for a particular τ_C can be computed as

$$S(f, \tau_C) = \int_{-\infty}^{\infty} R_x(\tau, \tau_C) e^{-j2\pi f \tau} d\tau = \frac{4A\tau_C}{1 + (2\pi f \tau_C)^2} \tag{5.41}$$

In a system with many simultaneous relaxation processes, τ_C can be viewed as a random variable with a continuous probability distribution. In such cases, the likelihood that $\tau_C = \rho$ is often described as a hyperbolic function

$$P(\rho) = \begin{cases} K/\rho, & \tau_1 \leq \rho \leq \tau_2 \\ 0, & \text{otherwise} \end{cases} \tag{5.42}$$

which takes nonzero values only on some interval $[\tau_1, \tau_2]$. This distribution reflects the assumption that processes with larger values of τ_C (i.e. processes

5.2. SELF-SIMILARITY IN THE TIME DOMAIN

which allow for long-term correlations between $x(t)$ and $x(t+\tau))$ are possible, but are less likely than those that have smaller relaxation times. Under such circumstances, the expected value of $S(f)$ takes the form

$$S(f) = E\left[S(f, \tau_C)\right] = \int_\rho S(f, \rho) P(\rho) d\rho = \frac{2AK}{\pi f} \quad (5.43)$$

in the frequency range

$$\frac{1}{2\pi\tau_2} \leq f \leq \frac{1}{2\pi\tau_1} \quad (5.44)$$

which corresponds to the $1/f$ spectrum that characterizes pink noise.

Since expression (5.39) represents a power law, one would expect to see some form of statistical self-similarity when white, brown and pink noise are scaled in the time domain. To establish whether this is indeed the case, it is helpful to first develop an intuitive idea of what this type of scaling entails. The easiest way to do that is to think of natural fluctuations as audible signals (since such a conversion is always possible). Scaling a random signal in time would then be equivalent to playing the recording faster or slower, while amplitude scaling would correspond to increasing or decreasing the volume.

In order to describe this process a bit more precisely, let us consider a random signal $x(t)$ which is defined on the interval [0 100], and the function $\hat{x}(t) = ax(t/b)$ obtained from it by scaling time by b and the amplitude by a. If we choose $b = 10$, this transformation is equivalent to taking the "slice" of $x(t)$ that corresponds to the interval [0 10], and "stretching" it to [0 100], with possible amplification (if $a > 1$) or attenuation (if $a < 1$).

How do different types of noise behave when they are scaled in this manner? When analyzing whether or not function $x(t)$ can be characterized as "self-similar", it makes no sense to look for a mapping that will transform $\hat{x}(t)$ into an exact copy of $x(t)$. Instead, we need to examine whether these two functions have the same *statistical properties* on interval [0 100]. It turns out that white noise meets this criterion perfectly, since its mean and variance are completely invariant to scaling transformations. As a result, we will always observe the same uncorrelated pattern (which is a lot like TV "static"), no matter what we do to such a signal.

Brownian noise is much less "robust" in that respect, since its statistical properties remain unchanged only for a special choice of scaling, which corresponds to $b = 2$ and $a = \sqrt{2}$. The diagrams shown in Figs. 5.5-5.7 illustrate this property, and show that the modified signal is quite different from the original if the amplitude is *not* increased by a factor of $\sqrt{2}$. In the case when $a = 1$, the signal becomes "flatter" when subjected to repeated scaling, while choosing $a = 2$ produces the opposite effect (in this case, the fluctuations become increasingly more pronounced).[1]

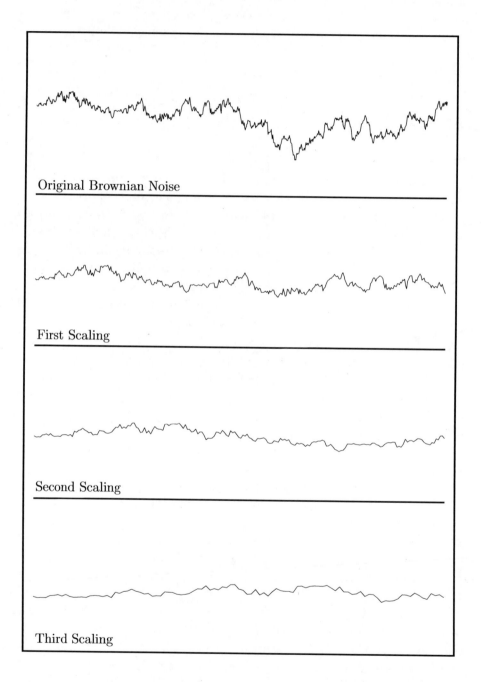

Figure 5.5: Scaled brown noise with $b = 2$ and $a = 1$.

5.2. SELF-SIMILARITY IN THE TIME DOMAIN

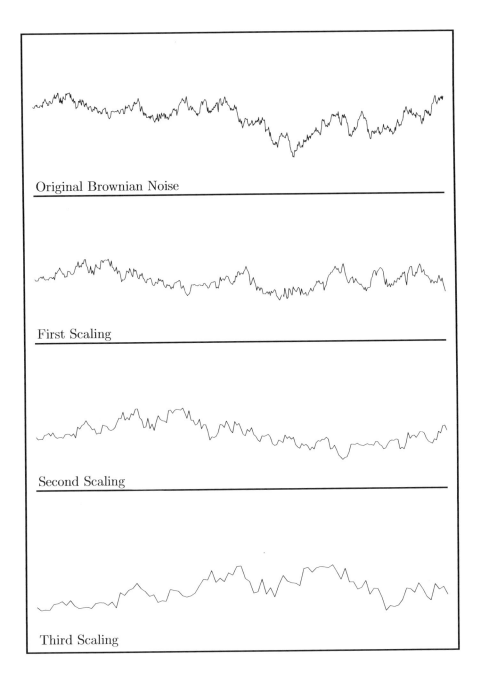

Figure 5.6: Scaled brown noise with $b = 2$ and $a = \sqrt{2}$.

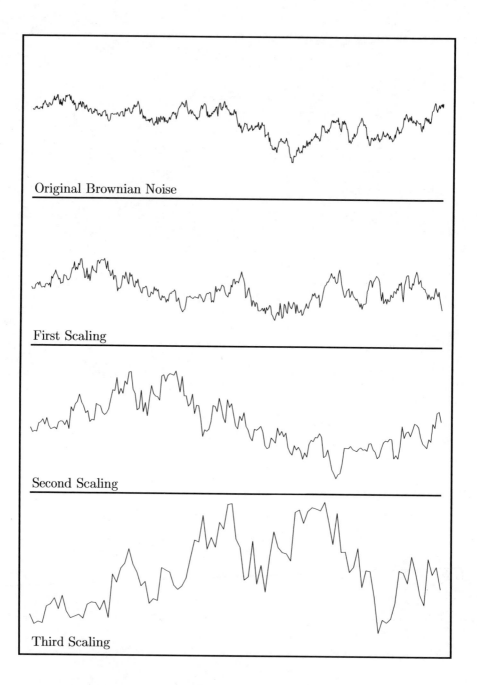

Figure 5.7: Scaled brown noise with $b = 2$ and $a = 2$.

Chapter 6

Fractals

In this chapter we will take a closer look at a special class of self-similar objects which are known as *fractals*. One of the things that makes fractals unusual is the fact that their "size" cannot be properly evaluated by standard measurement techniques (their "length", for example, depends on the precision of the "ruler" that is used, and is therefore an undefined quantity). Another difference between ordinary geometric forms and fractals has to do with the amount of information that is revealed as the measurement scale is decreased. An object like a square, for example, displays exactly the same features no matter how much we "zoom" into it, while a fractal constantly reveals new details on every level of magnification. Because of this property, fractals are said to be *scale-free* (in other words, there is no characteristic scale that corresponds to their smallest "interesting" feature).

6.1 Fractal Dimensions

Mathematical Fractals

Since conventional geometric concepts such as length and characteristic scale do not apply to fractals, the first question that we must address is how such objects can be meaningfully described. One way to approach this problem would be to recognize that fractals are often formed by recursive algorithms which allow us to relate the length of the ruler, l_n, and the number of rulers needed to cover the object in step n of the construction process (denoted $N(l_n)$). The term "ruler" should be understood rather broadly in this context - it could be a line segment of length l_n, a square of size $l_n \times l_n$, or any other geometric object that is characterized by a single linear dimension (the choice that is most appropriate is usually determined by the way the fractal is constructed).

When the relationship between $N(l_n)$ and l_n has the form of a power law for all n, the fractal is said to be *strictly self-similar* (which is consistent with the way this property was defined in the previous chapter). Laws of this sort are typically expressed as

$$N(l_n) = K l_n^{-D_S} \tag{6.1}$$

where K is a constant, and the exponent D_S is referred to as the *self-similarity dimension*. This number can be found either analytically (based on the rules that define the construction process), or by plotting $\log N(l_n)$ as a function of $\log(1/l_n)$ and determining the slope of the resulting line.

The following examples show how D_S can be computed in practice, and illustrate that the self-similarity dimension needn't necessarily be an integer (which is another feature that distinguishes fractals from "ordinary" geometric objects).

Example 6.1. The Cantor set is arguable the simplest fractal that can be formed using an iterative procedure. The basic idea behind this process is remarkably simple - we start with a line of unit length, and repeatedly transform it by removing the middle third of each segment (as shown in Fig. 6.1).

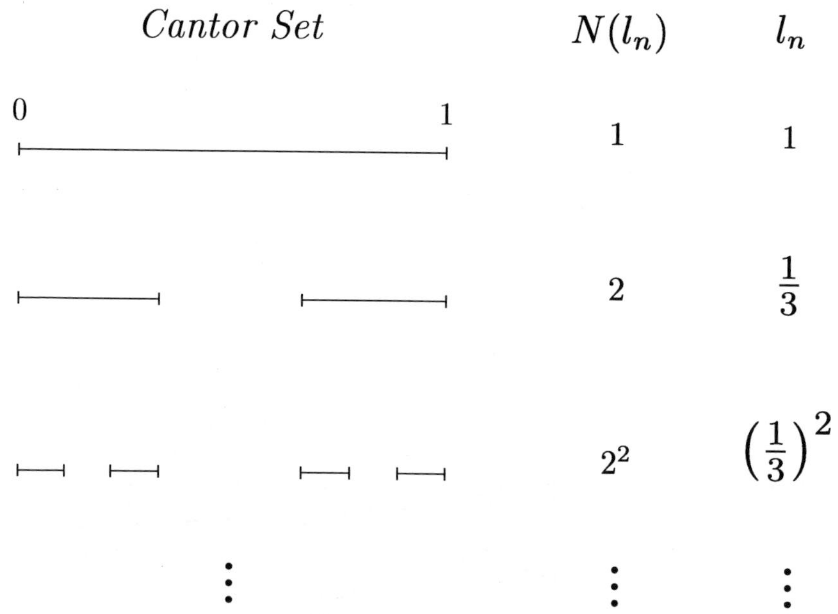

Cantor Set	$N(l_n)$	l_n
0 ——————————— 1	1	1
—— ——	2	$\frac{1}{3}$
— — — —	2^2	$\left(\frac{1}{3}\right)^2$
⋮	⋮	⋮

Figure 6.1: Construction of the Cantor set.

In order to determine the self-similarity dimension of the Cantor set, it suffices to observe that we need a single "ruler" of length $l_0 = 1$ to cover the set in Step 0, 2 "rulers" of length $l_1 = 1/3$ to do this in Step 1, 4 "rulers" of

6.1. FRACTAL DIMENSIONS

length $l_2 = 1/9$ in Step 2, and so on. Since this pattern remains unchanged throughout the process, it follows that in Step n we need 2^n "rulers" of size $l_n = (1/3)^n$ to cover the entire object. This allows us to compute D_S as

$$D_S = \frac{\log N(l_n)}{\log(1/l_n)} = \frac{\log 2^n}{\log 3^n} = 0.63 \tag{6.2}$$

How should we interpret such a result? The fact that the dimension of the Cantor set is a number between 0 and 1 suggests that this object is neither a collection of isolated points, nor a continuous curve. Instead, it is "something in between" (which is not a concept that Euclidean geometry is equipped to deal with).

Example 6.2. The construction of the Koch curve follows an iterative algorithm that is similar to the one described in Example 6.1, except for the fact that the removed "middle" segments are now replaced by two sides of an equilateral triangle. The first few steps of this procedure as shown in Fig. 6.2, which clearly indicates that we need 4^n "rulers" of size $l_n = (1/3)^n$ to cover the object in Step n. In view of that, the corresponding self-self similarity dimension can be calculated as

$$D_S = \frac{\log N(l_n)}{\log(1/l_n)} = \frac{\log 4^n}{\log 3^n} = 1.261 \tag{6.3}$$

It is important to note in this context that the length of the Koch curve is *undefined*, since

$$L_n = N(l_n) \cdot l_n = \left(\frac{4}{3}\right)^n \tag{6.4}$$

grows exponentially. What we measure will therefore vary dramatically with the size of the "ruler".

It is also worth mentioning that the Koch curve has a number of interesting variants, one of which is obtained by using a coin toss to decide whether the "peak" of each triangle faces "up" or "down". Figure 6.3 shows the first four steps of this process, which indicate that the resulting object looks much more like a natural form (such as a coastline, for example) than like a mathematically constructed fractal.

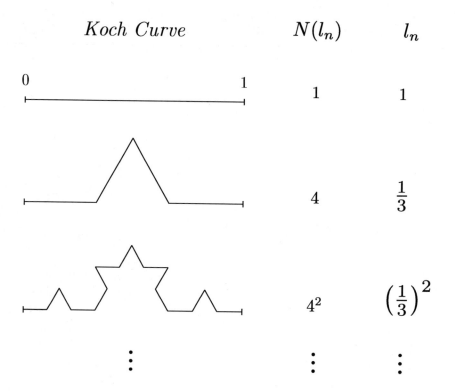

Figure 6.2: Construction of the Koch curve.

Example 6.3. The Minkowski curve is constructed using the procedure outlined in Fig. 6.4. In the first step, the unit segment is divided into four equal parts, and the middle two are replaced by the three sides of a square. If we choose the ruler size to be $l_1 = (1/4)$, it is easily seen that the number of "rulers" needed to cover this object will be $N(l_1) = 8$. In Step 2, we repeat this procedure on each of the four segments, which means that $l_2 = (1/4)^2$ is the appropriate ruler size. In that case we need $N(l_2) = 8^2$ such "rulers" to cover the entire object.

According to this algorithm, in Step n we will have $l_n = (1/4)^n$ and $N(l_n) = 8^n$, so the self-similarity dimension of the Minkowski curve can be computed as

$$D_S = \frac{\log N(l_n)}{\log(1/l_n)} = \frac{\log 8^n}{\log 4^n} = 1.5 \qquad (6.5)$$

As in the case of the Koch curve, the length of this object increases exponentially, since

$$L_n = N(l_n) \cdot l_n = \left(\frac{8}{4}\right)^n = 2^n \qquad (6.6)$$

6.1. FRACTAL DIMENSIONS

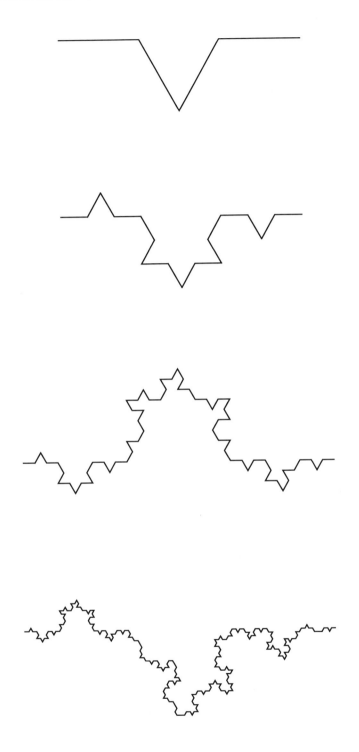

Figure 6.3: Construction of a "randomized" Koch curve.

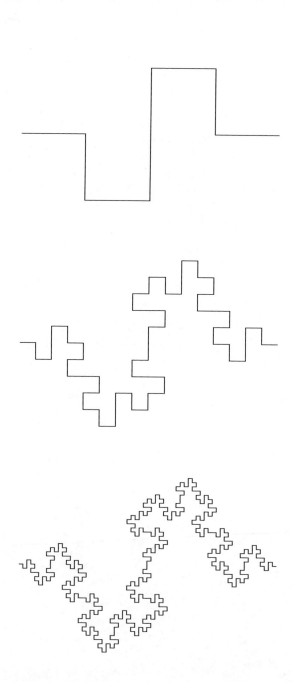

Figure 6.4: Construction of the Minkowski curve.

6.1. FRACTAL DIMENSIONS

Example 6.4. The Peano curve is somewhat different from the previous three fractals, since the number of segments increases rapidly from one step to the next. The first two stages of the construction process are shown in Figs. 6.5 and 6.6, indicating that $N(l_1) = 9$ and $N(l_2) = 9^2$ (the corresponding ruler sizes are $l_1 = 1/3$ and $l_2 = (1/3)^2$, respectively). From this, we can conclude that the appropriate ruler size in Step n is $l_n = (1/3)^n$, and that $N(l_n) = 9^n$.

If we substitute the obtained values for l_n and $N(l_n)$ into the expression for calculating D_S, we have that

$$D_S = \frac{\log N(l_n)}{\log(1/l_n)} = \frac{\log 9^n}{\log 3^n} = 2 \qquad (6.7)$$

This result may seem somewhat surprising, since it suggests that the Peano curve is actually a *two dimensional* object, although all of its intermediate stages consist of line segments. Such a situation, however, is not all that unusual in nature. The distribution of blood vessels in the body, for example, forms an extraordinarily intricate tree-like pattern which fills the three dimensional space it is embedded in. It turns out that this is a highly functional configuration, which optimizes the flow of blood in the body. Something similar can be said for the lungs as well, which fit into our chests very comfortably although their surface area exceeds $100 \ m^2$.

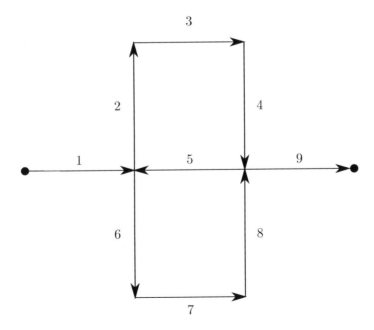

Figure 6.5: Step 1 in the construction of the Peano curve.

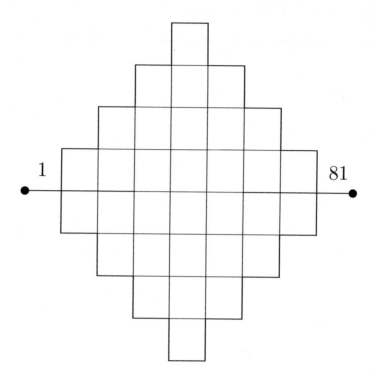

Figure 6.6: Step 2 in the construction of the Peano curve.

Example 6.5. In this example, we will consider a fractal that is known as the Sierpinski triangle. The construction of this object begins with a "filled" equilateral triangle with sides of unit length. If we partition this object in the manner shown in Fig. 6.7 and remove the middle triangle, we can cover the remaining figure with $N(l_1) = 3$ "rulers" (which, in this case, are equilateral triangle whose sides are of length $l_1 = 1/2$).

Repeating this operation on each shaded triangle produces the geometric form shown in Fig. 6.8, which can be covered by $N(l_2) = 3^2$ equilateral triangles with sides of length $l_2 = (1/2)^2$. If we continue this process indefinitely, we eventually obtain the fractal object in Fig. 6.9, whose visual appearance is quite striking.

The self-similarity dimension of this object (which is known as the Sierpinski triangle) can be easily computed if we observe that in the n-th step of the process we have $N(l_n) = (3)^n$ and $l_n = (1/2)^n$. Consequently, we obtain

$$D_S = \frac{\log N(l_n)}{\log(1/l_n)} = \frac{\log 3^n}{\log 2^n} = 1.58 \qquad (6.8)$$

6.1. FRACTAL DIMENSIONS

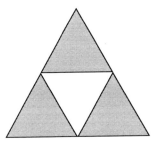

Figure 6.7: The first step in the construction of the Sierpinski triangle.

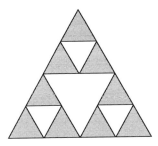

Figure 6.8: The second step in the construction of the Sierpinski triangle.

Figure 6.9: The Sierpinski triangle.

Example 6.6. The following example is interesting because it shows that we can construct a geometric object that conforms to a power law, but whose parts needn't necessarily produce an exact replica of the "whole" when they are magnified. The process begins by partitioning a unit square into 9 equal parts, and randomly selecting one that will be discarded (such squares are *not* shaded, as indicated in Fig. 6.10). If we use squares of dimension $1/3 \times 1/3$ as "rulers", it is not difficult to see that we need $N(l_1) = 8$ of them to cover the entire object in Step 1.

If we repeat this procedure on each of the shaded squares, after Step 2 we might obtain a pattern such as the one shown in Fig. 6.11 (I emphasize the word "might", since decisions about which squares should be discarded are made randomly). What is *not* uncertain, however, is the fact that we need $N(l_2) = 8^2$ "rulers" of size $1/9 \times 1/9$ to cover all the shaded areas.

Since the object obtained after n steps can be covered by $N(l_n) = 8^n$ "rulers" of size $(1/3)^n \times (1/3)^n$, we can conclude that the construction process conforms to a power law of the form

$$N(l_n) = l_n^{-D_S} \tag{6.9}$$

where

$$D_S = \frac{\log N(l_n)}{\log(1/l_n)} = \frac{\log 8^n}{\log 3^n} = 1.893 \tag{6.10}$$

Nevertheless, it is clear that magnifying one of the smaller parts by a factor of 3^n is not guaranteed to produce the pattern shown in Fig. 6.10. Indeed, since the boxes that are discarded are chosen arbitrarily, the most we can say is that there is a nonzero probability that one of the smaller parts will be an exact replica of the whole.

In evaluating the generality of the self-similarity dimension, it is important to keep in mind that this quantity can also be defined for objects that are *not* fractals. As an illustration, consider a line of length L, and a ruler of size s. In this case, the number of "rulers" needed to cover the entire object is obviously

$$N(s) = Ls^{-1} \tag{6.11}$$

which implies that $D_S = 1$. By the same token, a square of dimension $L \times L$ has a self-similarity dimension $D_S = 2$, since it can be covered with

$$N(s) = L^2 s^{-2} \tag{6.12}$$

smaller squares of size $s \times s$. This suggests that the definition of D_S is compatible with our intuitive notion of dimensionality, and is applicable to "ordinary" geometric objects as well.[2]

6.1. FRACTAL DIMENSIONS

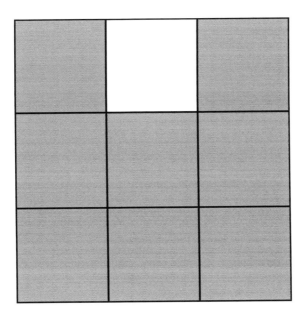

Figure 6.10: The object after a randomly chosen square is removed.

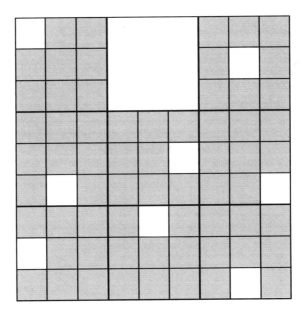

Figure 6.11: The second step in the construction process.

We should also point out that the notion of a fractal dimension is by no means limited to "ordinary" geometric objects and strictly self-similar fractals. To see how it can be extended to fractals that are *not* self-similar, let us consider an iterative construction process in which $N(l_n)$ and l_n are related as

$$N(l_n) = l_n^{-D(n)} \qquad (6.13)$$

where exponent $D(n)$ *varies* in each step. In this case, we cannot assign a meaningful self-similarity dimension to the resulting object, since it fails to conform to condition (6.1). That does not prevent us, however, from defining the quantity

$$D_C = \lim_{n \to \infty} \frac{\log(N(l_n))}{\log(1/l_n)} \qquad (6.14)$$

as its fractal dimension (if the limit exists, of course).

The difference between D_C (which is known as the *capacitive dimension*) and D_S is illustrated in Figs. 6.12 and 6.13, which show plots of $\log N(l_n)$ vs. $\log(1/l_n)$ for a strictly self-similar fractal, and for one that is not. In the former case, the plot is a straight line whose slope is D_S. The graph in Fig. 6.13, on the other hand, converges to a line only when $l_n \to 0$, so D_S can not be properly defined. The capacitive dimension D_C can, however, and we can identify it with the slope of the asymptote.

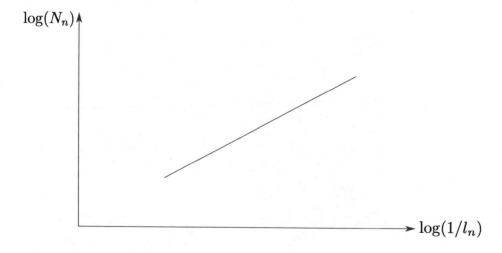

Figure 6.12: Function $\log(N_n)$ vs $\log(l_n)$ for a self-similar fractal.

6.1. FRACTAL DIMENSIONS

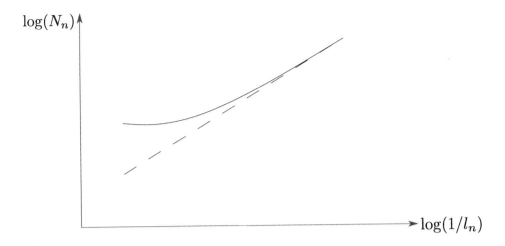

Figure 6.13: Function $\log(N_n)$ vs $\log(l_n)$ for a fractal that is not self-similar.

Irregular Fractals

The capacitive dimension is a useful concept in cases when we can establish an explicit relationship between $N(l_n)$ and l_n, but there are many situations where such a description is not available. Coastlines provide us with a typical example, since they meander randomly, and resemble mathematically generated fractals only in principle (but not in detail). As a result, we have no way of determining an analytic expression for $N(l_n)$.

One way to describe "irregular" forms of this sort would be to successively approximate the length of the object, and record how this quantity changes as the size of the "ruler" is reduced. In the case of coastlines we can do that pretty easily, using a detailed map and a compass. All that we need to do is identify each "ruler" length l_n with a different compass setting, and approximate the coastline by a collection of $N(l_n)$ line segments (in the manner shown in Figs. 6.14 and 6.15). Note that in this case the "rulers" do *not* cover the object completely - they simply connect points that belong to it.

Since the total length measured using the compass technique is $L_n = l_n \times N(l_n)$, we can plot $\log(L_n)$ as a function of $\log(1/l_n)$ and examine whether the graph can be approximated by a straight line with a slope d over a certain range of values for l_n (using the least squares method, for example). If this is possible, we can conclude that

$$L_n \approx K l_n^{-d} \tag{6.15}$$

which implies that

$$N(l_n) = L_n l_n^{-1} \approx K l_n^{-(d+1)} \tag{6.16}$$

This allows us to compute the *compass dimension* as $D_\alpha = d + 1$.

Figure 6.14: An approximation of the British coastline using line segments of equal length.

Figure 6.15: A more precise approximation of the British coastline.

6.2. MULTIFRACTALS

Since the compass dimension of natural fractals is usually not an integer, it would be interesting to see whether this changes when the approach outlined above is applied to regular geometric shapes. One way to test that would be to approximate a circle by a series of inscribed polygons (which is similar to the procedure that is used to determine D_α for coastlines). If we do so and plot $\log(L_n)$ versus $\log(1/l_n)$, we will obtain a *flat line*, which indicates that the overall length *does not* depend on the size of ruler. Since a flat line corresponds to $d = 0$, it follows that the compass dimension of a circle is $D_\alpha = 1$, which is consistent with the way its dimensionality is defined in Euclidean geometry.

An alternative to using a compass would be to partition the plane into *squares* of dimension $l_n \times l_n$, and interpret $N(l_n)$ as the number of boxes that contain segments of the object that we are interested in. Figure 6.16 shows how this works for the coastline of Britain. If we plot $\log N(l_n)$ as a function of $\log(1/l_n)$, we can identify the slope of the line as the *box dimension* of the object.[3-4] This technique is actually quite general, and allows us to evaluate the dimension of highly irregular planar fractals, including objects that have no clearly definable length (such as Jackson Pollock's paintings, for example).[5]

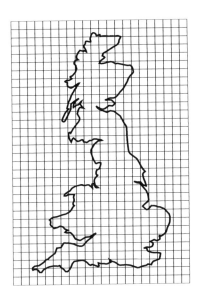

Figure 6.16: Computation of the box dimension for the British coastline.

6.2 Multifractals

Although most mathematically generated fractals have an easily computable capacitive dimension, describing such objects in terms of a single number can sometimes be inadequate. Fractals that have a complex substructure, for example, often need to be characterized by *several* dimensions, each of which corresponds to a different aspect of the object. Such "composite" geometric patterns are commonly referred to as *multifractals*, and their properties are illustrated by the following three examples.

Example 6.7. Consider the object shown in Fig. 6.17, where the line is of unit length and the squares have dimension $10^{-3} \times 10^{-3}$.

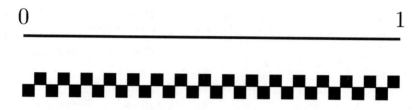

Figure 6.17: A composite geometric object.

If we use boxes of dimension $s \times s$ as "rulers" and pick larger values for s, the object will have all the characteristics of two parallel lines. This is easily seen from Fig. 6.18, which indicates that

$$N(s) = \frac{2}{s} \qquad (6.17)$$

and therefore

$$\log N(s) = \log 2 + \log(1/s) \qquad (6.18)$$

Equation (6.18) indicates that plotting $\log N(s)$ as a function of $\log(1/s)$ will produce a straight line whose slope is equal to 1.

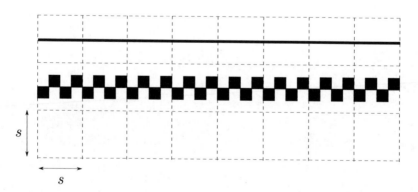

Figure 6.18: The object covered by boxes of dimension $s \times s$ (where $s > 10^{-3}$).

Expression (6.18) remains valid when the scale is decreased to $s = 10^{-3}$, as illustrated in Fig. 6.19. However, if s is decreased even further, a different

6.2. MULTIFRACTALS

structure begins to emerge. This is seen clearly in Fig. 6.20, which shows that each vertical "slice" contains $10^{-3}/s$ occupied boxes from the square, and a single additional box corresponding to the line.

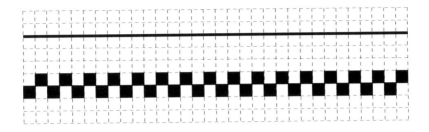

Figure 6.19: The object covered by boxes of dimension $10^{-3} \times 10^{-3}$.

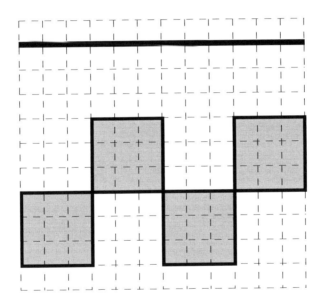

Figure 6.20: A small segment of the object covered by boxes of dimension $s \times s$ (where $s < 10^{-3}$).

Since the total number of "slices" is $1/s$, $N(s)$ takes the form

$$N(s) = \frac{1}{s} \cdot \left[\frac{10^{-3}}{s} + 1 \right] = \frac{10^{-3}}{s^2} + \frac{1}{s} \tag{6.19}$$

which becomes

$$\log N(s) = \log(1/s) + \log \left[\frac{10^{-3}}{s} + 1 \right] \tag{6.20}$$

after taking the logarithm of both sides. This expression indicates that the plot of $\log N(s)$ versus $\log(1/s)$ will begin to deviate from a straight line as s is decreased below 10^{-3}.

When $s \to 0$, the quadratic term in (6.19) becomes dominant, and $\log N(s)$ can be approximated as

$$\log N(s) \approx \log(10^{-3}/s^2) \tag{6.21}$$

This allows us to compute the capacitive dimension as

$$D_C = \lim_{s \to 0} \frac{\log(N(s))}{\log(1/s)} = \lim_{s \to 0} \frac{\log(10^{-3})}{\log(1/s)} + \lim_{s \to 0} 2 \frac{\log(1/s)}{\log(1/s)} = 2 \tag{6.22}$$

It is important to recognize, however, that if we were to characterize the object in Fig. 6.17 by this number alone, we would completely miss the fact that it consists of a surface *and* a line (since D_C captures only the *higher* of the two dimensions). On the other hand, if we analyze the *entire plot* in Fig. 6.21, it becomes apparent that this object needs to be characterized by *two* different dimensions (since the slope clearly changes from 1 to 2 as s becomes smaller).

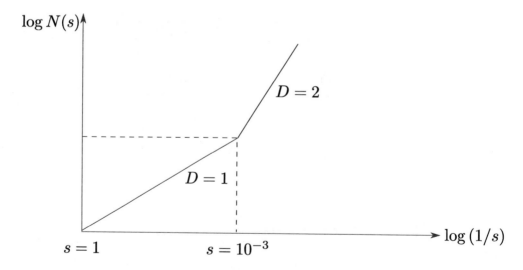

Figure 6.21: Asymptotic approximation of the relationship between $\log(N(s))$ and $\log(1/s)$.

Although this object is not a typical fractal (since the relevant dimensions are integers), it clearly illustrates the need for using different types of "rulers" - one for each pattern. If we were to do so, the line would produce $D_1 = 1$ when $s \to 0$, and the squares would yield $D_2 = 2$.

6.2. MULTIFRACTALS

Example 6.8. In this example, we will consider a combination of two "classical" mathematically generated fractals - the Cantor set and the Koch curve. Here it will be more convenient to use line segments of length $l_k = (1/3)^k$ instead of boxes, due to the nature of the underlying iterative algorithm. The construction process is illustrated in Fig. 6.22.

It is readily observed that in step n the number of rulers needed to cover the entire object is

$$N(l_n) = 2^n + 4^n = 4^n[1 + (1/2)^n] \quad (6.23)$$

As a result, we have

$$\frac{\log N(l_n)}{\log(1/l_n)} = \frac{\log 4^n}{\log 3^n} + \frac{\log[1+(1/2)^n]}{\log 3^n} = 1.261 + \frac{\log[1+(1/2)^n]}{\log 3^n} \quad (6.24)$$

and therefore

$$D_C = \lim_{n \to \infty} \frac{\log N(l_n)}{\log(1/l_n)} = \frac{\log 4}{\log 3} = 1.261 \quad (6.25)$$

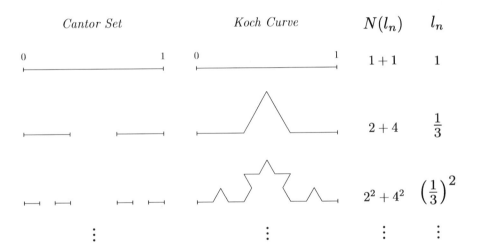

Figure 6.22: A combination of the Cantor set and the Koch curve.

Note that the result obtained in (6.25) is identical to the dimension of the Koch curve, and gives us no indication whatsoever that this object also contains a Cantor set. On the other hand, if we were to treat it as a *multifractal*, we could look at the two subsets separately, and compute a different dimension for each one. We could use $l_n = (1/3)^n$ for both subsets, obtaining

$$D_1 = \lim_{n \to \infty} \frac{\log N_1(l_n)}{\log(1/l_n)} = \frac{\log 2^n}{\log 3^n} = 0.63 \quad (6.26)$$

for the part that corresponds to the Cantor set, and

$$D_2 = \lim_{n\to\infty} \frac{\log N_2(l_n)}{\log(1/l_n)} = \frac{\log 4^n}{\log 3^n} = 1.261 \qquad (6.27)$$

for the part that corresponds to the Koch curve. This approach clearly reveals a substructure that is lost when we use a single dimension to characterize the fractal.

If we were to plot $\log N(l_n)$ as a function of $\log(1/l_n)$ (rather than focus exclusively on the limit when $n \to \infty$), we would clearly see that the object under examination is *not* a pure Koch curve, since it does not produce a straight line (it approaches a straight line *asymptotically* when $n \to \infty$).

Example 6.9. Let us consider a "composite" fractal made up of the standard Cantor set (formed using the "middle thirds" algorithm) and a variant of this procedure where we keep only 1/9 of the segment on each end (as shown in Fig. 6.23)

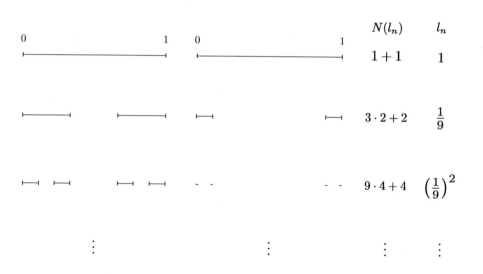

Figure 6.23: A fractal composed of two different types of Cantor sets.

It is not difficult to see that in this case an appropriate ruler size is $l_n = (1/9)^n$, and that the corresponding number of rulers needed to cover the object in step n is

$$N(l_n) = 6^n + 2^n = 6^n[1 + (1/3)^n] \qquad (6.28)$$

Consequently,

$$\frac{\log N(l_n)}{\log(1/l_n)} = \frac{\log 6^n}{\log 9^n} + \frac{\log[1 + (1/3)^n]}{\log 9^n} = 0.815 + \frac{\log[1 + (1/3)^n]}{\log 9^n} \qquad (6.29)$$

and
$$D_C = \lim_{n \to \infty} \frac{\log N(l_n)}{\log(1/l_n)} = \frac{\log 6}{\log 9} = 0.815 \quad (6.30)$$

It is important to recognize in this case that $D_C = 0.815$ tells us nothing about the two distinct substructures that make up the fractal, nor does it suggest that Cantor sets are involved in any way (recall that the capacitive dimension of the standard Cantor set is $D_C = 0.631$). However, a plot of $\log N(l_n)$ as a function of $\log(1/l_n)$ would show that this fractal is not strictly self-similar, since it approaches a straight line only when $n \to \infty$.

To see how the multifractal approach can help in this case, let us now treat the two subsets separately, using $l_n = (1/3)^n$ for the first one and $l_n = (1/9)^n$ for the second one. In that case, we have

$$D_1 = \lim_{n \to \infty} \frac{\log N_1(l_n)}{\log(1/l_n)} = \frac{\log 2^n}{\log 3^n} = 0.631 \quad (6.31)$$

for the part that corresponds to the standard Cantor set, and

$$D_2 = \lim_{n \to \infty} \frac{\log N_2(l_n)}{\log(1/l_n)} = \frac{\log 2^n}{\log 9^n} = 0.315 \quad (6.32)$$

for the other one. It is interesting to note in this context that $D_1 + D_2 = 0.946$, which is different from the dimension of the fractal as a whole (we found that to be $D = 0.815$).

6.3 Constructing Fractals of a Given Dimension

All of the "mathematical" fractals that we examined so far were constructed using well defined rules, which allowed us to establish a relationship between the ruler size l_n and the number of rulers needed to cover the object, $N(l_n)$. Based on this information, we could determine the capacitive dimension of the object using expression (6.14). In this section we will examine whether the procedure can be reversed, and whether we can systematically construct a fractal that has a *preassigned* capacitive dimension.

To see how this might be done, let us first consider an algorithm that was originally developed by Archimedes to calculate the area under a parabola. The procedure begins by drawing a line between the two bottom ends of the curve. We then identify the midpoint of this line, and connect it to the parabola with a vertical line, obtaining point $a(1)$ (as shown in Fig. 6.24)

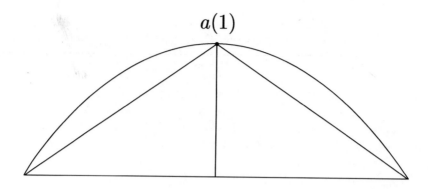

Figure 6.24: The first step in the approximation of a parabola.

The situation after the second stage is shown in Fig. 6.25. Here we connected the midpoints of each side of the triangle with the parabola, obtaining two additional points, $b(1)$ and $b(2)$. After one more step, we obtain *four* more points, $c(1)$, $c(2)$, $c(3)$ and $c(4)$, as shown in Fig. 6.26. It is not difficult to see that the total area of the triangles approximates the area under the parabola with increasingly greater accuracy as $n \to \infty$, and that 2^{n-1} new points are generated in Step n.

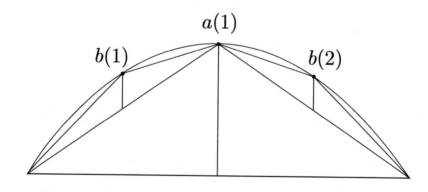

Figure 6.25: The second step in the approximation of a parabola.

Apart from providing an effective method for calculating the area under a parabola (without using integrals), this algorithm has another very interesting feature. Namely, it turns out that the heights of the vertical segments in any two successive steps satisfy the recursive relationship

$$h(n+1) = \frac{1}{4}h(n) \qquad (6.33)$$

6.3. CONSTRUCTING FRACTALS OF A GIVEN DIMENSION

We could therefore construct the parabola directly, by simply adding vertical segments to the midpoint of each side of the triangle (provided, of course, that the heights of these segments conform to expression (6.33)).

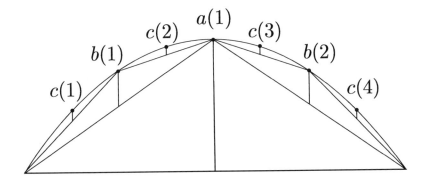

Figure 6.26: The third step in the approximation of a parabola.

German mathematician Georg Landsberg used this idea to generalize the algorithm of Archimedes by allowing the height of the vertical segments to vary as

$$h(n+1) = w \cdot h(n) \quad (6.34)$$

from one step to the next (where w is a preassigned number that belongs to the open interval $(0.5, 1)$).[6] The first three steps of such a procedure with $w = 0.75$ are shown in Fig. 6.27.

What makes this construction algorithm particularly interesting is the fact that the resulting object is guaranteed to be a *fractal* when w is chosen so that $0.5 < w < 1$. It can be shown that the box dimension of such an object is

$$D = 2 - |\log_2 w| \quad (6.35)$$

which means that we can construct fractals whose dimension is an *arbitrary* number between 1 and 2 by simply choosing an appropriate value for w. Doing this *exactly*, would, of course, require infinitely many steps, but we can obtain a pretty good approximation in a limited number of iterations.[7]

The midpoint displacement method can be further generalized by adding a random component to the construction algorithm. The simplest way to do this would be to use a coin toss, which would determine whether the vertical line will go *up* or *down* from the midpoint. We could formally express this modification as

$$h(n+1) = \rho(n) \cdot w \cdot h(n) \quad (6.36)$$

where $\rho(n)$ is a *discrete* random variable that takes on values -1 and 1. Figures 6.28-6.31 illustrate how such a curve can be constructed and what it might look like after the first four steps.

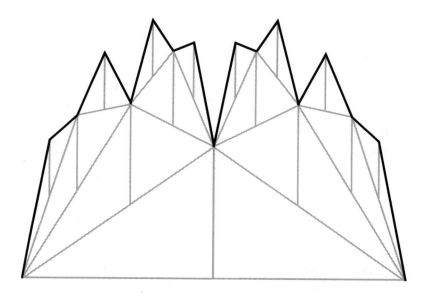

Figure 6.27: The first three steps of Landsberg's algorithm with $w = 0.75$.

A more sophisticated variant of this approach is based on a relationship of the form

$$h(n+1) = \rho \cdot w \cdot h(n) \qquad (6.37)$$

where $\rho \in [-1\ 1]$ is a *continuous* random variable with a Gaussian distribution. At each midpoint, a different randomly generated value of ρ is used (given the symmetry of the Gaussian function, positive and negative values for ρ are equally likely).

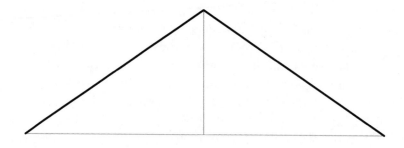

Figure 6.28: The first step of the randomized algorithm.

6.3. CONSTRUCTING FRACTALS OF A GIVEN DIMENSION 119

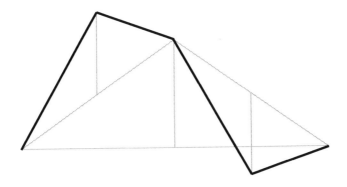

Figure 6.29: The second step of the randomized algorithm.

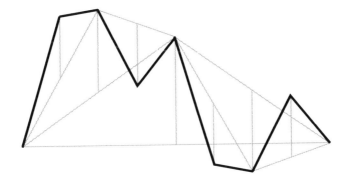

Figure 6.30: The third step of the randomized algorithm.

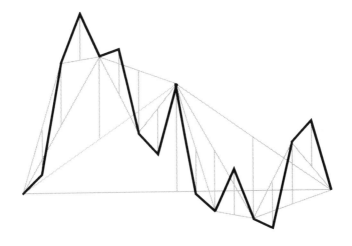

Figure 6.31: The fourth step of the randomized algorithm.

6.4 Fractals as Attractors

The fractals that we considered in Examples 6.1-6.5 were described in terms of an iterative process, in which the figure was systematically transformed in each step. It is possible, however, to adopt a somewhat different approach, and think of fractals as *attractors* of specially constructed mappings. To see what this means, let us once again consider the Koch curve which was described in Example 6.2. It is not difficult to show that each step in the construction of this figure can be viewed as a combination of the following four similarity transformations, which are schematically represented in Figs. 6.32-6.35.

Transformation 1. Scale set $A_0 = [0\ 1]$ by $1/3$.

Transformation 2. Scale set $A_0 = [0\ 1]$ by $1/3$, rotate it by $60°$, and shift by $1/3$ in the x-direction.

Transformation 3. Scale set $A_0 = [0\ 1]$ by $1/3$, rotate it by $-60°$, shift by $1/2$ in the x-direction, and by $\sqrt{3}/6$ in the y-direction.

Transformation 4. Scale set $A_0 = [0\ 1]$ by $1/3$, and shift it by $2/3$ in the x-direction.

Applying the *union* of these four operations to every point of set $A_0 = [0\ 1]$ produces a new set A_1, which represents the first step in the construction of the Koch curve. Note that in this process, each point $(x, y) \in A_0$ is mapped into *four* different points in A_1, which means that this is *not* a 1-to-1 mapping. We can formally describe such a transformation as

$$A_1 = W(A_0) \tag{6.38}$$

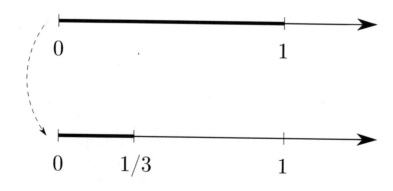

Figure 6.32: Schematic representation of Transformation 1.

6.4. FRACTALS AS ATTRACTORS

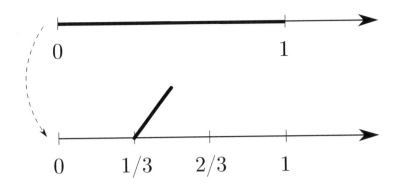

Figure 6.33: Schematic representation of Transformation 2.

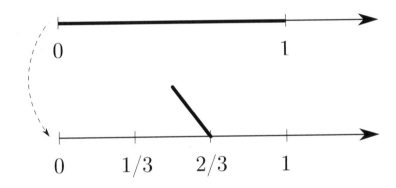

Figure 6.34: Schematic representation of Transformation 3.

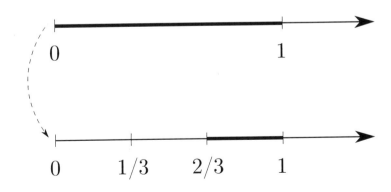

Figure 6.35: Schematic representation of Transformation 4.

where

$$W(A_0) = W_1(A_0) \cup W_2(A_0) \cup W_3(A_0) \cup W_4(A_0) \tag{6.39}$$

It is not difficult to show that $A_2 = W(A_1)$ represents the *second* step in the construction of the Koch curve (Figs. 6.36 - 6.39 show how this works). The same line of reasoning can be applied to all subsequent steps, so we can conclude that the iterative process

$$A_{n+1} = W(A_n) \tag{6.40}$$

produces a sequence of sets $\{A_0, A_1, \ldots, A_n, \ldots\}$ that correspond to different stages in the construction of the Koch curve.

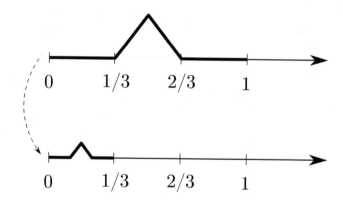

Figure 6.36: Transformation $W_1(A_1)$.

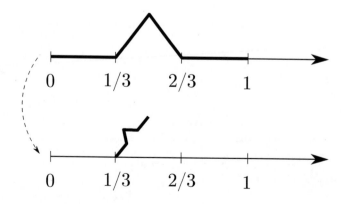

Figure 6.37: Transformation $W_2(A_1)$.

6.4. FRACTALS AS ATTRACTORS

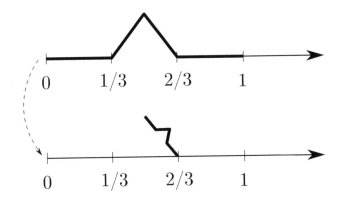

Figure 6.38: Transformation $W_3(A_1)$.

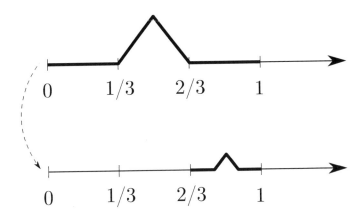

Figure 6.39: Transformation $W_4(A_1)$.

Although the process outlined above yields the same results as the method described in Example 6.2, the two are not equivalent. The iterative mapping in (6.40) is much more general, since it produces the Koch curve even if the initial set A_0 is *not* the unit interval (in fact, we obtain the same limiting object regardless of what geometric form we start from). As a result, we can think of the Koch curve as a *global attractor* for the iterative process described in (6.40). Note that this attractor (denoted A_∞) is also a *fixed point* for mapping W, since

$$A_\infty = W(A_\infty) \qquad (6.41)$$

This means (among other things) that applying transformation W to the Koch curve leaves it *unchanged*.

The idea of repeatedly applying a collection of similarity transformations to the same object is central to the notion of *iterated function systems* (IFS). Fig. 6.40 illustrates how this approach can be used to generate the Sierpinski triangle, which represents the attractor for the sequence

$$A_{n+1} = \tilde{W}(A_n) \tag{6.42}$$

where $\tilde{W}$ denotes the mapping that characterizes this particular fractal. Note that the same attractor is reached regardless of how the initial object A_0 is chosen.

Fractals as Fixed Points

We will conclude this section by demonstrating how one can rigorously prove that a fractal is the fixed point of some mapping W. It will be easiest to do this for the Cantor set, since the underlying iterative process is quite straightforward.

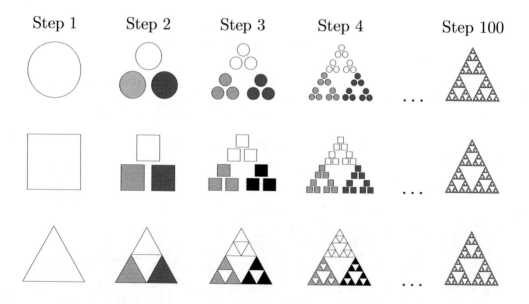

Figure 6.40: The Sierpinski triangle as an attractor for mapping $\tilde{W}$.

The mapping W for the Cantor set has the form

$$W(A) = W_1(A) \cup W_2(A) \tag{6.43}$$

6.4. FRACTALS AS ATTRACTORS

where
$$W_1(x) = 1/3x \qquad (6.44)$$

and
$$W_2(x) = 1/3x + 2/3 \qquad (6.45)$$

for any point x that belongs to set $A_0 = [0\ 1]$. The effects of these two transformations on A_0 are shown in Figs. 6.41 and 6.42, indicating that $A_1 = W(A_0)$ is indeed the first step in the construction of the Cantor set.

In order to show that the Cantor set is the fixed point for mapping W, we first need to prove the following simple lemma.

Lemma 6.1. Any point x that belongs to the Cantor set can be expressed as
$$x = a_1 \cdot 3^{-1} + a_2 \cdot 3^{-2} + a_3 \cdot 3^{-3} + \ldots \qquad (6.46)$$

in base 3 notation, where where coefficients a_i ($i = 1, 2, \ldots$) take values 0 or 2, but *not* 1.

Proof. In general, any number x that belongs to interval $[0\ 1]$ can be represented as
$$x = a_1 \cdot 3^{-1} + a_2 \cdot 3^{-2} + a_3 \cdot 3^{-3} + \ldots \qquad (6.47)$$

where the coefficients a_i ($i = 1, 2, \ldots$) can take values 0, 1 or 2. This expansion (which is usually given in its compact form $x = 0.a_1a_2a_3\ldots$) is referred to as the *base 3 representation* of x.

As in the case of decimals or binary numbers, there is an inherent non-uniqueness in this expansion that needs to be resolved before proceeding any further. To see this more clearly, we should observe that a number such as $x = 1/3$ can be represented in two equivalent ways: as
$$x = 0.1 \qquad (6.48)$$

or as
$$x = 0.022222\ldots = 0.0\bar{2} \qquad (6.49)$$

Note that the latter form corresponds to the expansion
$$x = 2 \cdot (3^{-2} + 3^{-3} + \ldots) = 2 \cdot 3^{-2} \cdot \sum_{n=0}^{\infty} \left(\frac{1}{3}\right)^n \qquad (6.50)$$

in which the infinite sum is a geometric series that converges to $3/2$. As a result, we have that
$$x = 2 \cdot 3^{-2} \cdot \frac{3}{2} = \frac{1}{3} \qquad (6.51)$$

In order to avoid this ambiguity, in the following we will adopt the representation shown in (6.48).

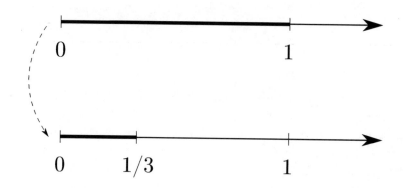

Figure 6.41: Schematic representation of transformation $W_1(A_0)$.

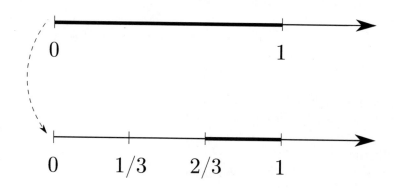

Figure 6.42: Schematic representation of transformation $W_2(A_0)$.

Let us now consider the first step in the construction of the Cantor set. It is not difficult to see that removing the "middle third" amounts to eliminating all numbers from interval [0 1] which have $a_1 = 1$ (i.e., a 1 in the first decimal place). Indeed, any number

$$x = 0.0a_2a_3a_4\ldots \qquad (6.52)$$

satisfies

$$x \leq 0.022\bar{2} = \frac{1}{3} \qquad (6.53)$$

while numbers of the form

$$x = 0.2a_2a_3a_4\ldots = 2 \cdot 3^{-1} + \varepsilon \qquad (6.54)$$

necessarily belong to the interval [2/3 1].

6.4. FRACTALS AS ATTRACTORS

In Step 2 of the "middle thirds" algorithm, one of the intervals that we eliminate is (1/9 2/9), whose elements have the form

$$x = 0.01a_3a_4a_5\ldots \tag{6.55}$$

This is obviously equivalent to removing all numbers with a 1 in the *second* decimal place. Proceeding in this manner, it is not difficult to see that the only numbers that will "survive" are those that have *only zeros and twos* in their base 3 expansion. **Q.E.D.**

We can now prove the following theorem.

Theorem 6.1. Let C denote the Cantor set, and let W represent the iterative mapping that is used to construct it. Then, C is a fixed point for mapping W.

Proof. In order to show that $C = W(C)$, let us first assume that $x \in C$ and apply operator W to this point. Since W represents the union of mappings W_1 and W_2, we need to show that *both* of these mappings transform x into an element of the Cantor set. In the case of W_1, we have

$$W_1(x) = 0 \cdot 3^{-1} + a_1 3^{-2} + a_2 3^{-3} + \ldots \tag{6.56}$$

since this mapping scales each point by 1/3. Recalling that all a_i's are either 0 or 2 (by virtue of Lemma 6.1), it is clear that $W_1(x) \in C$. A similar argument can be applied to W_2, which amounts to scaling x by 1/3 and then adding 2/3 to the result. This operation produces

$$W_2(x) = 0 \cdot 3^{-1} + a_1 3^{-2} + a_2 3^{-3} + \ldots + (2 \cdot 3^{-1}) \tag{6.57}$$

or equivalently

$$W_2(x) = 2 \cdot 3^{-1} + a_1 3^{-2} + a_2 3^{-3} + \ldots \tag{6.58}$$

It is obvious that this number belong to the Cantor set as well, since all the coefficients are either zeros or twos.

Now that we have established that $W(C) \subset C$, we must also show that $C \subset W(C)$. This means that for any point $y \in C$, there must be some $x \in C$ such that $y = W(x)$. Note that W maps point x into *two* different points, $x' = W_1(x)$ and $x'' = W_2(x)$, so it suffices to show that *one* of these corresponds to the given y.

To verify that this is indeed the case, let us consider an arbitrary element y of the Cantor set, whose form must necessarily be

$$y = a_1 3^{-1} + a_2 3^{-2} + a_3 3^{-3} + \ldots \tag{6.59}$$

If a_1 happens to be 0, we have

$$y = 0 \cdot 3^{-1} + a_2 3^{-2} + a_3 3^{-3} + \ldots \qquad (6.60)$$

Setting

$$x = a_2 3^{-1} + a_3 3^{-2} + a_4 3^{-3} + \ldots \qquad (6.61)$$

and applying W_1, we obtain

$$W_1(x) = 0 \cdot 3^{-1} + a_2 3^{-2} + a_3 3^{-3} + \ldots \qquad (6.62)$$

which is precisely how y was defined.

If, on the other hand, y is chosen so that $a_1 = 2$, we have

$$y = 2 \cdot 3^{-1} + a_2 3^{-2} + a_3 3^{-3} + \ldots \qquad (6.63)$$

When W_2 is applied to the same x as in (6.61), we obtain

$$W_2(x) = 0 \cdot 3^{-1} + a_2 3^{-2} + a_3 3^{-3} + \ldots + (2 \cdot 3^{-1}) \qquad (6.64)$$

or equivalently

$$W_2(x) = 2 \cdot 3^{-1} + a_2 3^{-2} + a_3 3^{-3} + \ldots \qquad (6.65)$$

which is obviously the same number as y. We can therefore conclude that $C \subset W(C)$, which is what we had to show. **Q.E.D.**

6.5 Generalized Dimensions

Fractal structures can also arise in the context of dynamic systems, both continuous and discrete. The following example provides an interesting illustration of this possibility.

Example 6.10. Consider the discrete dynamic system

$$\begin{aligned} x_{k+1} &= 0.5\,[x_k + x_R] \\ y_{k+1} &= 0.5\,[y_k + y_R] \end{aligned} \qquad (6.66)$$

where x_R and y_R are *randomly* chosen in each step from a set of three preassigned pairs: (x_1, y_1), (x_2, y_2) and (x_3, y_3). If we interpret these pairs as the vertices of a triangle and pick an arbitrary initial condition (x_0, y_0) in its interior, sequence (6.66) will always converge to the attractor shown in Fig. 6.43. It is not difficult to see that this attractor is actually a Sierpinski triangle, which is a fractal whose self-similarity dimension is 1.58 (as demonstrated in Example 6.5). What is particularly interesting in this case is the fact that such an attractor exists although the system is *not* fully deterministic.

6.5. GENERALIZED DIMENSIONS

Figure 6.43: The attractor for the sequence described in (6.66).

On first glance, it would appear that the box counting method is a natural way to determine the fractal dimension of such attractors. Indeed, for each choice of s, we would simply need to record how many boxes have been "visited" by the system trajectory after a sufficiently large number of iterations has been performed. It turns out, however, that determining the box dimension in this way entails a number of serious difficulties.

To get a sense for why such calculations can be problematic, consider a general second order discrete system of the form

$$\begin{aligned} x_{k+1} &= f(x_k, y_k) \\ y_{k+1} &= g(x_k, y_k) \end{aligned} \qquad (6.67)$$

Given an initial condition (x_0, y_0), such a system will typically exhibit transient behavior for the first 50-100 iterations, after which the solution becomes sufficiently close to the attractor. When that point is reached, we can partition the state space into boxes of size $s \times s$ and examine which ones are populated and which ones are not. If point (x_k, y_k) appears in a box that is not already populated, we need to increase the box count $N(s)$ by one (since a new box has just been visited). Otherwise, we do nothing, and proceed to the next iterate.

If we choose to apply this strategy, we will initially have boxes that are visited multiple times, and $N(s)$ will increase as the box size is reduced (since the heavily populated boxes will turn into multiple less heavily populated ones, thus increasing the overall count). As s continues to decrease, however, the grid will become finer, and the number of visited boxes could remain unchanged for prolonged periods of time.

It is not particularly difficult to see why this is so - when s becomes small enough, each point on the attractor will have *its own box*, and $N(s)$ will not change any more. Before we reach that critical point, however, there will be a range of values of s for which $N(s)$ increases *very slowly*. This suggests that we may have to perform a huge number of iterations before a new box is visited for small values of s. It also shows that there is no reliable criterion for ending the process, since the fact that the system hasn't visited a new box in as many as a million iterations doesn't guarantee that this won't happen in the very next one.

Another shortcoming of this method is that it tells us nothing about *how many times* a box has been visited - it just distinguishes between boxes that are populated and those that are not. As a result, the box dimension provides no insight into the dynamic behavior of the system. Such considerations have led mathematicians to explore alternative definitions of dimensionality, which can capture *both* the geometric characteristics of the object *and* the process that gives rise to it.

In order to see what such a definition might look like, let us consider a general second order discrete system of the form

$$x_{k+1} = f(x_k) \qquad (6.68)$$

where x_k represents a vector of dimension 2×1. If this system has a fractal attractor, the sequence of points $\{x_0, x_1, \ldots, x_n, \ldots\}$ will converge to it after the transients have gone away.

Suppose now that we partition the state space in which the attractor is located into boxes of size $s \times s$, and number these boxes. The probability that an iterate will "land" in box k could then be calculated as

$$p_k(s) = \lim_{n \to \infty} \frac{F_k(n, s)}{n} \qquad (6.69)$$

where $F_k(n, s)$ represents the number of times box k has been visited after n iterations. Because this number *depends on the size of the box*, we must treat p_k as a function of s.

Introducing a probabilistic element into the computation of dimensionality reflects the fact that there is some uncertainty about which box a given iterate x_i will visit. Since this resembles the kind of uncertainty that we encounter in communication systems, it makes sense to apply some of the concepts and terminology that are used in information theory.

6.5. GENERALIZED DIMENSIONS

With that in mind, we will define the *information content* associated with box k as
$$I_k(s) = -\log p_k(s) \tag{6.70}$$

and interpret the *expected information content*
$$I(s) = -\sum_{k=1}^{N(s)} p_k(s) \log p_k(s) \tag{6.71}$$

of a box as the *information entropy* of the attractor. The term $N(s)$ in expression (6.71) denotes the *total* number of boxes of size s, including those that are empty. Note, however, that this is equivalent to taking the sum over *all populated boxes*, since the empty ones will have $p_k = 0$.[8]

In order to relate the information entropy of an attractor to its dimensionality, let us assume that the dependency of $I(s)$ on s can be approximated as
$$I(s) = I_0 + D_I \log_2\left(\frac{1}{s}\right) \tag{6.72}$$

(which happens to be realistic in many cases). Under such circumstances, D_I will satisfy
$$D_I = \lim_{s \to 0} \frac{I(s)}{\log(1/s)} \tag{6.73}$$

which looks very much like the expression we used to compute the *capacitive dimension* for fractals (except that $\log N(s)$ is now replaced by $I(s)$). Because of this similarity, D_I is commonly referred to as the *information dimension* of the attractor.

In order to establish a more formal connection between D_I and the capacitive dimension (and to create room for other possible definitions), it is helpful to introduce the notion of *generalized information entropy*. This quantity was defined by Hungarian mathematician Alfréd Rényi as
$$I_q(s) = \frac{1}{1-q} \log \chi(q, s) \tag{6.74}$$

where
$$\chi(q, s) = \sum_{k=1}^{N(s)} p_k^q(s) \tag{6.75}$$

The term $p_k^q(s)$ in (6.75) represents the probability that q iterates will "land" in box k, so $\chi(q, s)$ can be interpreted as the *probability that a box of size $s \times s$ will be visited q times*.

Using expression (6.74), we can now define the corresponding *generalized dimension* as

$$D_q = \lim_{s \to 0} \frac{I_q(s)}{\log(1/s)} = \frac{1}{1-q} \cdot \lim_{s \to 0} \frac{\log \chi(q,s)}{\log(1/s)} \qquad (6.76)$$

To get a better understanding of what expression (6.76) really means, it is helpful to examine how $I_q(s)$ and D_q vary as functions of q. In the following, we will focus on cases when $q = 0$, 1 and 2, since these three values have particularly useful interpretations.

CASE 1. For $q = 0$, we have

$$I_0(s) = \log \sum_{k=1}^{N(s)} p_k^0(s) = \log N(s) \qquad (6.77)$$

and

$$D_0 = \lim_{s \to 0} \frac{I_0(s)}{\log(1/s)} = \lim_{s \to 0} \frac{\log N(s)}{\log(1/s)} \qquad (6.78)$$

This is obviously the same expression as (6.14) (where s plays the role of l_n), so we can conclude that D_0 is equivalent to the *capacitive dimension*.

CASE 2. The case when $q = 1$ is a bit trickier, since the denominator in expression (6.76) becomes zero. This difficulty can be avoided by using L'Hospital's rule, which eventually produces

$$I_1(s) = -\sum_{k=1}^{N(s)} p_k(s) \log p_k(s) \qquad (6.79)$$

after some straightforward algebraic manipulations. It is not difficult to see that this precisely matches the definition of information entropy introduced in (6.71) (and in Section 1.2 as well). The corresponding dimension D_1 then becomes

$$D_1 = \lim_{s \to 0} \frac{I_1(s)}{\log(1/s)} \qquad (6.80)$$

which is consistent with the way D_I was defined in (6.73). It therefore makes sense to refer to D_1 as the *information dimension*.

CASE 3. For $q = 2$, we have

$$I_2(s) = -\log \sum_{k=1}^{N(s)} p_k^2(s) \qquad (6.81)$$

which implies

$$D_2 = \lim_{s \to 0} \frac{I_2(s)}{\log(1/s)} = -\lim_{s \to 0} \frac{\log \sum_{k=1}^{N(s)} p_k^2(s)}{\log(1/s)} \qquad (6.82)$$

6.5. GENERALIZED DIMENSIONS

The quantity defined by expression (6.82) is commonly referred to as the *correlation dimension*, since

$$P(s) = \sum_{k=1}^{N(s)} p_k^2(s) \qquad (6.83)$$

represents the probability that *a pair of iterates* will appear in the *same* box of size $s \times s$. Since the distance between all iterates in the same box is smaller than s, (4.83) can also be interpreted as the *probability that the distance between two iterates is smaller than s*. This is convenient because such a probability can be estimated quite easily.

Chapter 7

Notes and References

7.1 Notes for Chapters 5-6

1. In principle, it is possible to generate an entire family of random signals that behave similarly to Brownian noise when exposed to scaling. The statistical properties of these signals (which are known as *fractional Brownian noise*) remain unchanged when $b = 2$ and $a = 2^H$, where H is a number between 0 and 1. It is not difficult to see that ordinary Brownian noise corresponds to the special case when $H = 0.5$.

2. We should note that there is a subtle difference between the way D_S is computed for "ordinary" objects and fractals. In expressions (6.11) and (6.12) the ruler length s was be chosen *arbitrarily*, which is usually not possible for fractals. The Cantor set and the Koch curve, for example, have a "natural" choice of ruler size, whose length is $l_n = (1/3)^n$ in the n-th step of the construction process. The power law from which D_S is derived holds for this particular choice of l_n, but not in general.

3. It is important to recognize that most physical objects do not allow us to reduce l_n below a certain value, so we can realistically speak only of fractal dimensions over a *certain range of scales*. This restriction becomes quite obvious in the case of coastlines, since measurements make no sense beyond a certain threshold (we cannot account for patterns formed by grains of sand, for example).

4. We should also keep in mind that the box dimension and the compass dimension are *not* identical. In the case of the British coastline, the compass dimension is found to be 1.36, while the value obtained by counting boxes produces 1.31. The difference between the two can be attributed

to the fact that the box dimension is more general, since it can capture irregularities such as islands, which are not taken into account when only length is measured.

5. See: Richard Taylor, "Fractal expressionism – where art meets science", in *Art and Complexity*, J. Casti and A. Karlqvist (Eds.), Elsevier Science, 2003.

6. See: Benoit Mandelbrot, "Midpoint displacement and systematic fractals: The Takagi fractal curve, its kin, and the related systems", in *The Science of Fractal Images*, H. Peitgen and D. Saupe (Eds.), Springer, 1988.

7. It is usually convenient to express w as $w = 2^{-H}$, where H ($0 < H < 1$) is known as the *Hurst exponent*. In that case, we have

$$D = 2 - H \qquad (7.1)$$

which relates this exponent to the box dimension of the curve.

8. According to classical information theory, $I(s)$ can also be viewed as the expected minimal number of Yes/No questions needed to determine which box of size $s \times s$ an iterate will visit. We can say, in other words, that expression (6.71) represents the amount of information needed to locate a point on the attractor with an accuracy of s.

7.2 Further Reading

1. Benoit Mandelbrot, *The Fractal Geometry of Nature*, W. H. Freeman, 1982.

2. Joseph McCauley, *Chaos, Dynamics and Fractals*, Cambridge University Press, 1994.

3. Philip Iannaccone and Mustafa Khokha, *Fractal Geometry in Biological Systems*, CRC Press, 1996.

4. Jean-Francois Gouyet, *Physics and Fractal Structures*, Springer, 1996.

7.2. FURTHER READING

5. T. Gregory Dewey, *Fractals in Molecular Biophysics*, Oxford University Press, 1998.

6. Benoit Mandelbrot, *Multifractals and 1/f Noise: Wild Self-Affinity in Physics*, Springer, 1999.

7. Peitgen, Jürgens and Saupe, *Chaos and Fractals: New Frontiers of Science*, Springer, 2004.

8. Manfred Schroeder, *Fractals, Chaos, Power Laws*, Dover Publications, 2009.

9. Kenneth Falconer, *Fractal Geometry: Mathematical Foundations and Applications*, Wiley, 2014.

Part III

Complexity and Self-Organization

Chapter 8

Nonlinearity, Chaos and Catastrophes

In the three chapters that follow, we will examine how simple rules and equations can give rise to elaborate spatiotemporal patterns and highly complex dynamic behavior. This might sound rather surprising to those who are not experts in the field, since our intuition tells us that complexity and simplicity are polar opposites. We also tend to think of complex structures as the result of numerous gradual changes that take place over a long period of time (as in the case of biological evolution). We will see, however, that both of these assumptions can be misleading, and will show that order in nature can emerge abruptly, and for no apparent "reason". We will also demonstrate that random perturbations can play a constructive role in this process (although we usually think of them of them as events that disrupt order).

To get a sense for how something like that is possible, we will first consider nonlinear phenomena such as chaos and catastrophes, which give rise to complex dynamics although the underlying models are very simple. We will then expand our discussion to cellular automata and random Boolean networks, which also exhibit this type of behavior (albeit in a different way). We will conclude with a brief description of how certain types of systems can spontaneously self-organize into a state where order and disorder can coexist. We will see that such "critical" states represent a potential source of complexity, since they allow for the emergence of new and qualitatively different forms.

8.1 Nonlinear Systems

A nonlinear dynamic system is generally described by a set of first order differential equations

$$\dot{x} = f(x, t) \qquad (8.1)$$

where $x(t) = [x_1(t)\ x_2(t)\ \ldots\ x_n(t)]^T$ represents the *state vector*, and $f(x,t) = [f_1(x,t)\ f_2(x,t)\ \ldots\ f_n(x,t)]^T$ is a nonlinear function. If function f depends only on x we say that the system is *autonomous*, and equation (8.1) reduces to

$$\dot{x} = f(x) \tag{8.2}$$

When studying the dynamic behavior of nonlinear systems of the form (8.2), one of the things that we are most interested in are their *constant* solutions (which are typically denoted by x^e). Any such solution must satisfy

$$f(x^e) = 0 \tag{8.3}$$

by definition, and is referred to as an *equilibrium* of the system. Equilibrium x^e is an *attractor* if the solutions of equation (8.2) converge to x^e when $t \to \infty$ for a range of initial conditions. If this property holds for *all* initial conditions, the attractor is said to be *global*.

A simple way to determine whether or not a given equilibrium is attractive is to *linearize* the system using the Taylor series expansion. For first order systems, this amounts to approximating the right hand side of the equation as

$$f(x) \approx f(x^e) + f'(x^e)(x - x^e) \tag{8.4}$$

and recognizing that the first term disappears by virtue of equation (8.3). This allows us to rewrite the original system as

$$\dot{x} = f'(x^e)(x - x^e) \tag{8.5}$$

in the neighborhood of x^e (where this approximation is valid). If we now introduce a new variable $y = x - x^e$, it is not difficult to see that (8.5) can be equivalently expressed as

$$\dot{y} = f'(x^e)y \tag{8.6}$$

since $\dot{y} = \dot{x}$.

This strategy applies to higher order system as well, for which the Taylor series expansion takes the form

$$f(x) \approx f(x^e) + J(x^e)(x - x^e) \tag{8.7}$$

In this case the role of the derivative is played by matrix $J(x^e)$, which represents the Jacobian

$$J = \begin{bmatrix} \partial f_1/\partial x_1 & \cdots & \partial f_1/\partial x_n \\ \vdots & & \vdots \\ \partial f_n/\partial x_1 & \cdots & \partial f_n/\partial x_n \end{bmatrix} \tag{8.8}$$

8.1. NONLINEAR SYSTEMS

evaluated at $x = x^e$. If we once again define a new variable $y = x - x^e$, the linearized system becomes

$$\dot{y} = J(x^e)y \qquad (8.9)$$

and its stability properties can be determined by analyzing the eigenvalues of $J(x^e)$.[1]

The following two examples illustrate how this approach works in practice.

Example 8.1. Let us consider a simple first order system

$$\dot{x} = x^2 + 2x \qquad (8.10)$$

which has a quadratic nonlinearity. Since the equation

$$0 = x^2 + 2x \qquad (8.11)$$

has solutions $x^e = 0$ and $x^e = -2$, it follows that this system has *two* different equilibria. In this particular case, the derivative at x^e has the form

$$f'(x^e) = 2x^e + 2 \qquad (8.12)$$

so linearization around $x^e = 0$ yields

$$\dot{y} = 2y \qquad (8.13)$$

The solutions of this equation have the form

$$y(t) = y(0)e^{2t} \qquad (8.14)$$

which means that $y(t)$ moves away from the equilibrium no matter how close we are to $x^e = 0$ at time $t = 0$ (we refer to such equilibria as *repellers*).

Linearization around $x^e = -2$ yields

$$\dot{y} = -2y \qquad (8.15)$$

whose solutions have the form

$$y(t) = y(0)e^{-2t} \qquad (8.16)$$

These solutions obviously converge to $y = 0$ when $t \to \infty$, so it follows that $x^e = -2$ is an *attractor* (since $y = x - x^e$).[2]

Example 8.2. To illustrate how this type of analysis works when the system description involves multiple equations, let us consider the so called predator-prey model

$$\begin{aligned} \dot{x}_1 &= -3x_1 + 4x_1^2 - 0.5x_1x_2 - x_1^3 \\ \dot{x}_2 &= -2.1x_2 + x_1x_2 \end{aligned} \qquad (8.17)$$

which is commonly used to study the dynamics of ecosystems. Our first step will be to determine the equilibria of this system, which requires solving the following pair of algebraic equations:

$$-3x_1 + 4x_1^2 - 0.5x_1x_2 - x_1^3 = 0$$
$$-2.1x_2 + x_1x_2 = 0$$
(8.18)

In a typical nonlinear system we would have to do this numerically, an there would be no guarantee that all possible solutions could be identified. In this case, however, the problem can be solved analytically, by observing that condition

$$x_2(x_1 - 2.1) = 0$$
(8.19)

allows for two possible cases.

CASE 1. $x_2 \neq 0$, which implies that $x_1 = 2.1$ and $x_2 = 1.98$ is the unique solution.

CASE 2. $x_2 = 0$, which implies that x_1 can be determined by solving equation

$$-3x_1 + 4x_1^2 - x_1^3 = 0$$
(8.20)

This is a simple third order polynomial with roots 0, 1 and 3.

Based on the above analysis, we can conclude that the predator-prey model has *exactly* four equilibria

$$x^e = \begin{bmatrix} 0 \\ 0 \end{bmatrix}; \quad x^e = \begin{bmatrix} 1 \\ 0 \end{bmatrix}; \quad x^e = \begin{bmatrix} 3 \\ 0 \end{bmatrix}; \quad x^e = \begin{bmatrix} 2.10 \\ 1.98 \end{bmatrix}$$
(8.21)

To evaluate their stability, we need to linearize the system in the manner described in equation (8.9). If all the eigenvalues of $J(x^e)$ happen to be in the left half of the complex plane, the corresponding equilibrium x^e will be an attractor.

The Jacobian of system (8.17) has the general form

$$J(x) = \begin{bmatrix} (-3 + 8x_1 - 0.5x_2 - 3x_1^2) & -0.5x_1 \\ x_2 & -2.1 + x_1 \end{bmatrix}$$
(8.22)

which leads to the following four scenarios:

(i) For $x^e = [0 \quad 0]^T$, we have

$$J(x^e) = \begin{bmatrix} -3 & 0 \\ 0 & -2.1 \end{bmatrix}$$
(8.23)

In this case, the eigenvalues of $J(x^e)$ are $\lambda_1 = -3$ and $\lambda_2 = -2.1$, indicating that $x^e = [0 \quad 0]^T$ is an *attractor*.

8.1. NONLINEAR SYSTEMS

(ii) For $x^e = [1 \ 0]^T$, we have

$$J(x^e) = \begin{bmatrix} 2 & -0.5 \\ 0 & -1.1 \end{bmatrix} \tag{8.24}$$

and the eigenvalues of the linearized system are $\lambda_1 = 2$ and $\lambda_2 = -1.1$. This indicates that $x^e = [1 \ 0]^T$ is a *repeller*.

(iii) For $x^e = [3 \ 0]^T$, we have

$$J(x^e) = \begin{bmatrix} -6 & -1.5 \\ 0 & 0.9 \end{bmatrix} \tag{8.25}$$

and the eigenvalues of the linearized system are $\lambda_1 = -6$ and $\lambda_2 = 0.9$. This indicates that $x^e = [3 \ 0]^T$ is a *repeller*.

(iv) For $x^e = [2.1 \ 1.98]^T$, we have

$$J(x^e) = \begin{bmatrix} -0.42 & -1.05 \\ 1.98 & 0 \end{bmatrix} \tag{8.26}$$

and the eigenvalues of the linearized system are $\lambda_{1,2} = -0.21 \pm j1.43$. This indicates that $x^e = [2.1 \ 1.98]^T$ is an *attractor*.

Although linearization allows us to conclude that two of the four equilibria are locally attractive, we should keep in mind that this pertains only to solutions that originate in the immediate neighborhood of these points. Determining how other solutions behave can be a difficult task, even for relatively simple systems such as the predator-prey model. One way to approach the problem is to perform multiple simulations, and then plot each trajectory in the $x_1 - x_2$ plane (which is commonly referred as the *phase space*). A collection of these trajectories represents the *phase portrait* of the system, and is shown in Fig. 8.1 (the solid dots indicate the initial conditions). Fig. 8.1 suggests that there are entire regions in the phase space from which solutions are attracted to particular equilibria (such regions are known as *basins of attraction*).

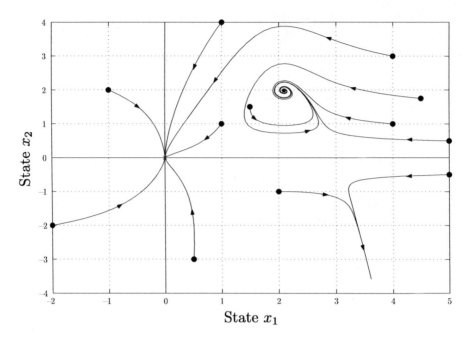

Figure 8.1: Regions of attraction for the two stable equilibria.

8.1.1 Types of Attractors in Nonlinear Systems

In the two examples that we considered so far, the trajectories were attracted to (or repelled by) the system equilibria. It turns out, however, that in some nonlinear systems trajectories can also be attracted by other "objects" in the phase space. In the following, we will identify three such objects, in increasing order of complexity.

Limit Cycles

The system shown in equation (8.27) is known as the van der Pol oscillator.

$$\dot{x}_1 = x_2$$
$$\dot{x}_2 = (1 - x_1^2)x_2 - x_1 \tag{8.27}$$

This system has a unique equilibrium $x^e = [0\ 0]^T$, and its Jacobian has the form

$$J(x) = \begin{bmatrix} 0 & 1 \\ -(2x_1 x_2 + 1) & 1 - x_1^2 \end{bmatrix} \tag{8.28}$$

Since

$$J(x^e) = \begin{bmatrix} 0 & 1 \\ -1 & 1 \end{bmatrix} \tag{8.29}$$

8.1. NONLINEAR SYSTEMS

has eigenvalues $\lambda_{1,2} = 0.5 \pm j0.866$, we can conclude that x^e is a *repeller*, and that solutions originating in the neighborhood of this point diverge from it.

It is important to recognize, however, that linear approximations cannot tell us what happens to these solutions once they move away from the equilibrium. The easiest way to evaluate their long term behavior is to solve equation (8.27) numerically for several different initial conditions (some close to the equilibrium, and some further away). A typical solution of this system is shown in Figs. 8.2 and 8.3, which indicate that $x_1(t)$ and $x_2(t)$ eventually settle into a periodic steady state. This steady state trajectory is referred to as a *limit cycle*, and represents a legitimate attractor for the system.

To see the attractive nature of the limit cycle more clearly, in Fig. 8.4 we provide a phase portrait of the system. In this diagram, the limit cycle corresponds to a *closed curve* to which the various different trajectories are drawn as $t \to \infty$.

Quasi-Periodic Attractors

Before we introduce the next type of attractor, it is first necessary to define what is meant by *quasi-periodic* dynamic behavior. The simplest example would be the function

$$f(t) = \cos \omega_1 t + \cos \omega_2 t \tag{8.30}$$

which is a sum of two sinusoids with different frequencies. Such a function may or may not be periodic, depending on how we choose ω_1 and ω_2. To see why this is so, we should recall that a function $f(t)$ will be periodic if condition

$$f(t+T) = f(t) \tag{8.31}$$

holds for some T. It is easily verified that this will be the case when

$$\begin{aligned}\omega_1 T &= 2p\pi \\ \omega_2 T &= 2q\pi\end{aligned} \tag{8.32}$$

for some integers p and q, or equivalently if

$$\frac{\omega_1}{\omega_2} = \frac{p}{q} \tag{8.33}$$

When condition (8.33) holds, we say that frequencies ω_1 and ω_2 are *commensurate*, and $f(t)$ will be periodic with

$$\omega_{eq} = \frac{2\pi}{T} = \frac{\omega_1}{p} = \frac{\omega_2}{q} \tag{8.34}$$

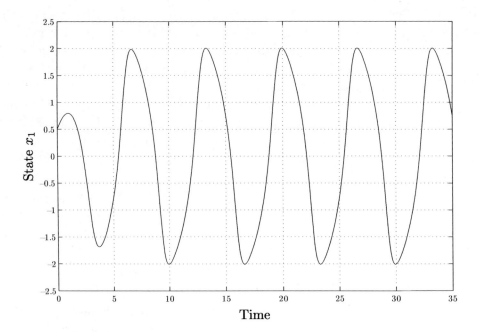

Figure 8.2: Evolution of state x_1.

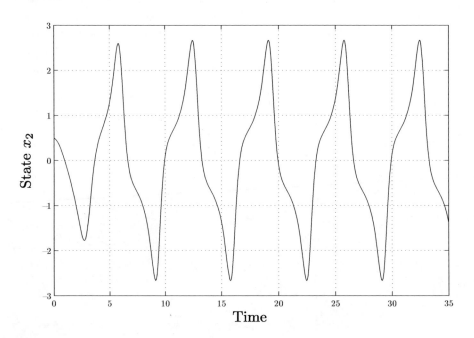

Figure 8.3: Evolution of state x_2.

8.1. NONLINEAR SYSTEMS

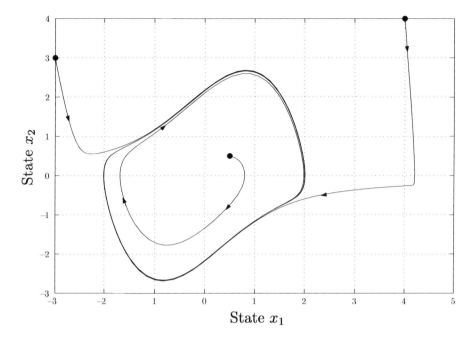

Figure 8.4: The attractive limit cycle in state space.

If ω_1 and ω_2 are *not* commensurate, $f(t)$ is clearly not periodic, but it can nevertheless be represented as a sum of two independent periodic functions. Because of this property, we say that $f(t)$ is *quasi-periodic*.

An example of a nonlinear system with a quasi-periodic attractor is the forced van der Pol oscillator

$$\begin{aligned} \dot{x}_1 &= x_2 \\ \dot{x}_2 &= (1 - x_1^2)x_2 - x_1 + 0.5\cos 1.1t \end{aligned} \qquad (8.35)$$

This is a *non-autonomous* system, which differs from the one in (8.27) only in the sinusoidal forcing term. The temporal evolution of $x_1(t)$ and $x_2(t)$ is shown in Figs. 8.5 and 8.6, which suggest that the steady state solutions of equation (8.35) are *not* periodic. The phase portrait for this system (which is provided in Fig. 8.7) indicates, however, that this system has a *quasi-periodic attractor*. It is important to note that this attractor is a two dimensional geometric object in the phase space, and is therefore more complex than the one dimensional closed curves that characterize limit cycles.

Strange Attractors

"Strange" attractors represent the most complicated type of attractor that can arise in a nonlinear system. The solutions associated with it appear to be completely random despite the fact that they are generated by a set of well defined deterministic equations. This unexpected (and thoroughly counterintuitive) property blurs the line between randomness and predictability, and suggests that simple rules can sometimes give rise to highly complex behavior.

The first system in which such behavior was observed was discovered by meteorologist Edward Lorenz in 1963.[3] The model that he studied had the form of a third order differential equation

$$\dot{x}_1 = -10x_1 + 10x_2$$
$$\dot{x}_2 = -x_1 x_3 + 28x_1 - x_2 \qquad (8.36)$$
$$\dot{x}_3 = x_1 x_2 - (8/3)x_3$$

which seems rather straightforward and unexceptional. It turns out, however, that its solutions exhibit completely random behavior in the time domain (as illustrated in Fig. 8.8).

What is even more surprising is the fact that this system has a "strange" attractor, whose geometric form is provided in Fig. 8.9 (this object happens to be a *fractal*, whose correlation dimension is slightly higher than 2).

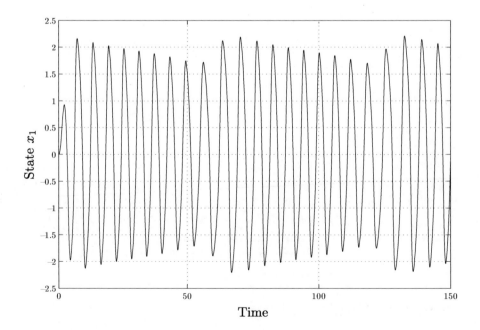

Figure 8.5: Evolution of state x_1.

8.1. NONLINEAR SYSTEMS

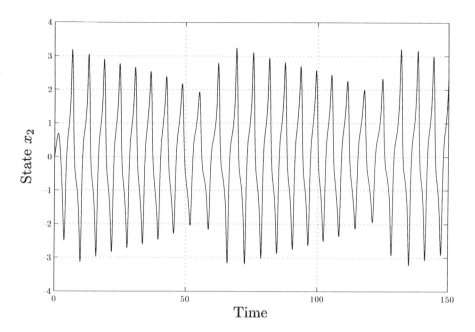

Figure 8.6: Evolution of state x_2.

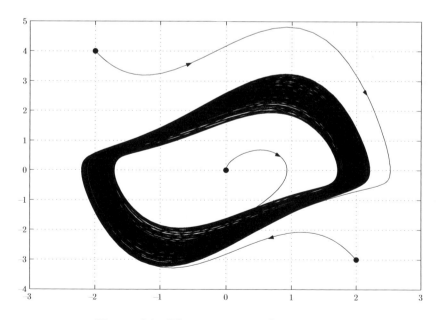

Figure 8.7: The quasi-periodic attractor.

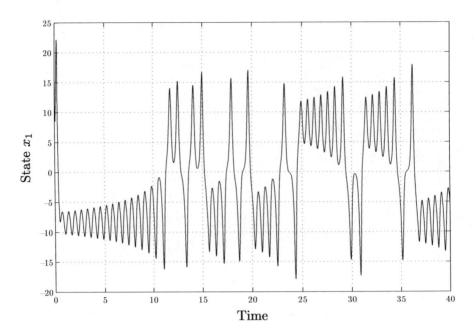

Figure 8.8: Evolution of state x_1.

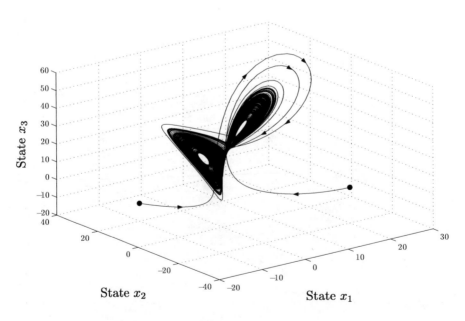

Figure 8.9: The strange attractor.

8.1. NONLINEAR SYSTEMS

The realization that such attractors exist and that there could be some kind of "hidden order" behind trajectories that show no signs of regularity is one of the most striking discoveries of chaos theory. What we have learned about such systems in recent years suggests that this sort of behavior is not unusual in nature, and is closely related to the emergence of complex dynamic patterns (both spatial and temporal).

8.1.2 Hypersensitivity

Sensitivity to Initial Conditions

Simple nonlinear models such as the Lorenz equation have a number of other interesting characteristics, one of which is their extreme sensitivity to parametric variations and changes in initial conditions. The latter property is illustrated in Fig. 8.10, which shows two solutions of system [4]

$$\begin{aligned} \dot{x}_1 &= 9x_2 - 9h(x_1) \\ \dot{x}_2 &= x_1 - x_2 + x_3 \\ \dot{x}_3 &= -(100/7)x_2 \end{aligned} \quad (8.37)$$

where $h(x_1)$ is a piecewise linear function which is defined as

$$h(x_1) = \begin{cases} (2/7)x_1 - 3/7, & x_1 \geq 1 \\ -(1/7)x_1, & -1 < x_1 < 1 \\ (2/7)x_1 + 3/7, & x_1 \leq -1 \end{cases} \quad (8.38)$$

Fig. 8.10 clearly indicates that any resemblance between the two solutions disappears after a sufficiently long time, despite the fact that their initial conditions differ by only 0.01% (for better contrast, one of the trajectories is displayed as black and the other one as gray).

The extraordinary sensitivity of system (8.37) to changes in initial conditions has a number of important consequences, one of which is that its long term behavior is virtually unpredictable. This is clearly not something that one would expect to see in a deterministic model. In order to gain some insight into the origins of this phenomenon, let us consider a general first order system of the form

$$x_{n+1} = f(x_n) \quad (8.39)$$

Although its mathematical description is quite simple, the dynamic behavior of such a system will allow us to illustrate how two initially indistinguishable solutions can eventually become very different.

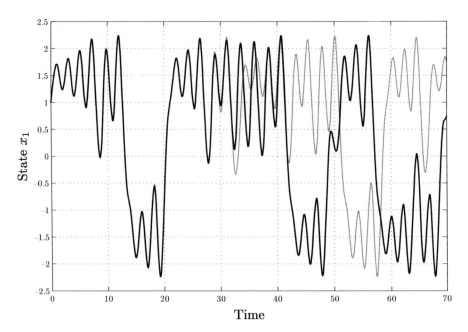

Figure 8.10: Evolution of two initially indistinguishable solutions.

To see how that can be done, let us assume that initial condition x_0 is perturbed by some amount δx_0. If δx_0 is sufficiently small, the value obtained after one iteration can be approximated as

$$x_1 + \delta x_1 = f(x_0 + \delta x_0) \approx f(x_0) + f'(x_0)\delta x_0 \tag{8.40}$$

Observing that the unperturbed system satisfies

$$x_1 = f(x_0) \tag{8.41}$$

expression (8.40) reduces to

$$\delta x_1 \approx f'(x_0)\delta x_0 \tag{8.42}$$

after the appropriate terms are canceled.

Proceeding in a similar manner, in the second step we have

$$x_2 + \delta x_2 = f(x_1 + \delta x_1) \approx f(x_1) + f'(x_1)\delta x_1 \tag{8.43}$$

which can be rewritten as

$$\delta x_2 \approx f'(x_1)\delta x_1 = f'(x_1)f'(x_0)\delta x_0 \tag{8.44}$$

8.1. NONLINEAR SYSTEMS

using (8.42) and the fact that $x_2 = f(x_1)$. If the perturbations remain small we can use this approach repeatedly, and estimate the difference between the two solutions as

$$\delta x_n \approx \prod_{i=0}^{n-1} f'(x_i) \delta x_0 \tag{8.45}$$

after n steps.

Expression (8.45) can be equivalently rewritten as

$$\ln\left(\frac{\delta x_n}{\delta x_0}\right) \approx \sum_{i=0}^{n-1} \ln\left[f'(x_i)\right] \tag{8.46}$$

after dividing both sides by δx_0 and taking the logarithm. If n is a sufficiently large number, we can further approximate the term on the right hand side of equation (8.46) as

$$\sum_{i=0}^{n-1} \ln\left[f'(x_i)\right] \approx n\lambda \tag{8.47}$$

where

$$\lambda = \lim_{n \to \infty} \frac{1}{n} \sum_{i=0}^{n-1} \ln\left[f'(x_i)\right] \tag{8.48}$$

represents the *Lyapunov exponent* for this system. Substituting (8.47) into (8.46), we now have that

$$\delta x_n \approx e^{n\lambda} \delta x_0 \tag{8.49}$$

which indicates that δx_n increases exponentially when $\lambda > 0$. From this, we can conclude that hypersensitivity with respect to initial conditions is a realistic possibility, and that such dynamic behavior is bound to arise in systems that have a positive Lyapunov exponent.[5]

Sensitivity to Parametric Variations

A rather different sort of hypersensitivity arises in systems whose models contain one or more parameters. As an illustration, let us consider the system

$$\begin{aligned} \dot{x}_1 &= p[x_2 - h(x_1)] \\ \dot{x}_2 &= x_1 - x_2 + x_3 \\ \dot{x}_3 &= -16 x_2 \end{aligned} \tag{8.50}$$

where $h(x_1)$ is the function described in equation (8.38) and p is a parameter that can vary continuously. The following three scenarios illustrate the impact of such variations on the system dynamics.

CASE 1. For $p = 8.8$ the system has a periodic solution with a *single* amplitude. This is evident from Fig. 8.11, which shows the evolution of state $x_1(t)$.

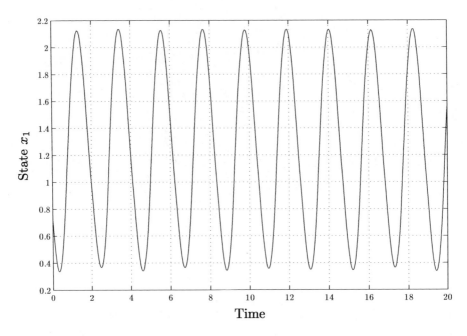

Figure 8.11: Periodic solution with a single amplitude.

CASE 2. When $p = 9.1$ (which differs from the previous value by only 3.5%), the solution is still periodic, but now has *two* distinct amplitudes. This phenomenon (which is known as *period doubling*) is illustrated in Fig. 8.12.

CASE 3. For $p = 9.3$ (which represents an even smaller change in parameter p), the system exhibits aperiodic behavior, whose general pattern is shown in Fig. 8.13.[6]

The trajectories shown in Figs. 8.11-8.13 suggest that the dynamic behavior of the system experiences qualitative changes as the parameter approaches certain critical values. These values (which are known as *bifurcation points*) are intimately related to the emergence of complexity and self-organization in nature. The following simple examples illustrate what might happen when such a point is reached.

Example 8.3. Consider the system

$$\dot{x} = p - x^2 \tag{8.51}$$

8.1. NONLINEAR SYSTEMS

where p is a scalar parameter. This system has no equilibrium when $p < 0$, since equation

$$p - x^2 = 0 \tag{8.52}$$

has no real solutions. For $p = 0$, there is a unique equilibrium at $x = 0$, and for $p > 0$ there are two equilibria corresponding to $\pm\sqrt{p}$.

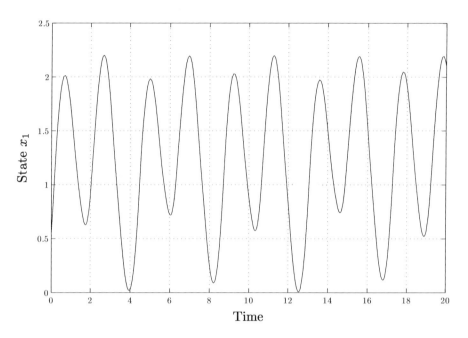

Figure 8.12: Periodic solution with two amplitudes.

The analysis outlined above is best summarized by plotting the system equilibria as a function of p, in the manner shown in Fig. 8.14. Such a plot (which is known as a *bifurcation diagram*) helps us pinpoint the "critical" parameter values for which the system behavior changes abruptly. It also provides some insight into the stability properties of the equilibria, since it distinguishes between those that are attractive and those that are not (the latter type of equilibrium is indicated by dashed lines).

In this particular case, $x^e = \sqrt{p}$ represents the "stable branch" of the bifurcation diagram, since linearizing the system around this point produces

$$\dot{y} = -2\sqrt{p}\,y \tag{8.53}$$

The other branch is unstable, since linearization around $x^e = -\sqrt{p}$ corresponds to

$$\dot{y} = 2\sqrt{p}\,y \tag{8.54}$$

158　　　CHAPTER 8.　NONLINEARITY, CHAOS AND CATASTROPHES

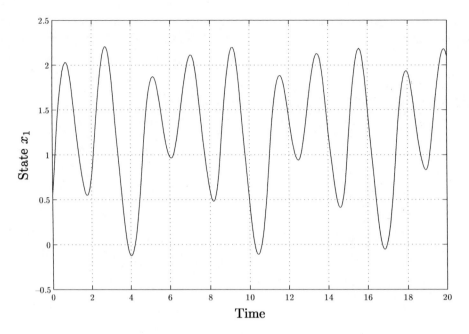

Figure 8.13: An aperiodic solution that corresponds to chaotic dynamics.

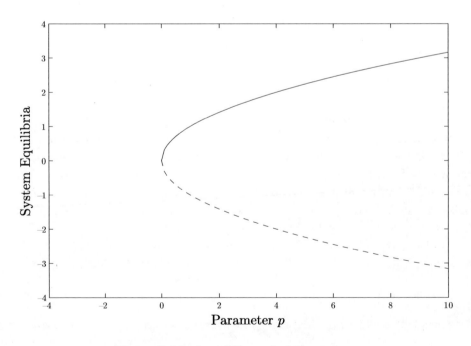

Figure 8.14: A saddle node bifurcation.

8.1. NONLINEAR SYSTEMS

It is obvious from Fig. 8.14 that this system undergoes a qualitative change for $p = 0$, when two new equilibria emerge. We can therefore say that $p = 0$ represents a bifurcation point. Since one of the new equilibria is always stable and the other is not, this type of bifurcation is commonly referred to as a *saddle-node*.

Example 8.4. Consider the system

$$\dot{x} = x(p - x^2) \tag{8.55}$$

For $p < 0$, the only equilibrium is $x = 0$, while for $p > 0$ there are two more, corresponding to $\pm\sqrt{p}$. The Jacobian associated with system (8.55) has the form

$$J(x) = p - 3x^2 \tag{8.56}$$

which gives rise to the following three scenarios:

(*i*) For $x^e = 0$, we have that $J(x^e) = p$, and the the linearized system becomes

$$\dot{y} = py \tag{8.57}$$

Consequently, we can conclude that this equilibrium is stable when $p < 0$, and unstable when $p > 0$.

(*ii*) For $x^e = \sqrt{p}$, the linearized system takes the form

$$\dot{y} = -2py \tag{8.58}$$

where $y = x - x^e$. Since this equilibrium exists only for $p > 0$, it is guaranteed to be stable.

(*iii*) For $x^e = -\sqrt{p}$, the linearized system is exactly the same as (8.58), and the same analysis applies. As a result, this equilibrium is stable for all $p > 0$.

The three possibilities outlined above are summarized in the bifurcation diagram shown in Fig. 8.15. From this figure, it is clear that $p = 0$ represents a bifurcation point for the system. This type of bifurcation (where two new *stable* equilibria emerge) is known as a *pitchfork*.

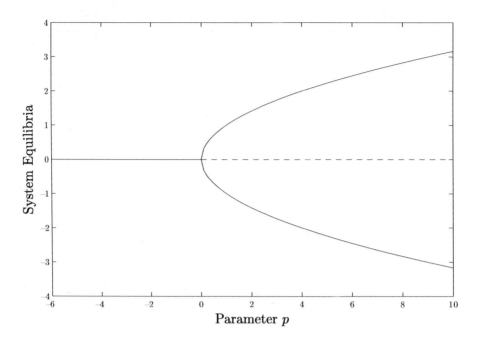

Figure 8.15: A pitchfork bifurcation.

Example 8.5. The system

$$\dot{x} = x(p - x) \tag{8.59}$$

has two equilibria for all values of p, located at $x^e = 0$ and $x^e = p$, respectively. If we linearize equation (8.59) around these equilibria, we have the following situation:

(i) For $x^e = 0$, the linearized system is

$$\dot{y} = py \tag{8.60}$$

Consequently, this equilibrium is stable when $p < 0$, and unstable when $p > 0$.

(ii) For $x^e = p$, the linearized system is

$$\dot{y} = -py \tag{8.61}$$

This equilibrium is stable when $p > 0$, and unstable when $p < 0$.

The bifurcation diagram for this system is shown in Fig. 8.16. This figure indicates that $p = 0$ is once again a bifurcation point, but the nature of the qualitative change is different from the previous two examples. In this case no

8.1. NONLINEAR SYSTEMS

new equilibria are created, and what we have instead is an *exchange of stability* between the the existing ones (such a bifurcation is said to be *transcritical*).

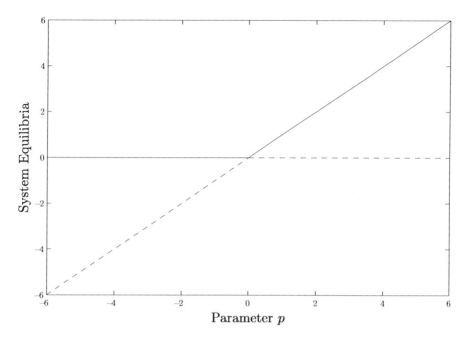

Figure 8.16: A transcritical bifurcation.

Our final example in this section demonstrates that bifurcations needn't necessarily be limited to equilibria. In the system that we will examine next, a limit cycle emerges for a particular parameter value and replaces the equilibrium as the system attractor (this type of transition is referred to in the literature as a *Hopf bifurcation*).

Example 8.6. Consider the system

$$\begin{aligned} \dot{x}_1 &= p_1 x_1 - p_2 x_2 - (x_1^2 + x_2^2) x_1 \\ \dot{x}_2 &= p_2 x_1 + p_1 x_2 - (x_1^2 + x_2^2) x_2 \end{aligned} \quad (8.62)$$

which contains two parameters, p_1 and p_2. It is easily verified that this system has an equilibrium at the origin, and that the corresponding Jacobian is

$$J(x^e) = \begin{bmatrix} p_1 & -p_2 \\ p_2 & p_1 \end{bmatrix} \quad (8.63)$$

The eigenvalues of $J(x^e)$ are $\lambda_{1,2} = p_1 \pm j p_2$, which means that this equilibrium is stable for $p_1 < 0$ and unstable for $p_1 > 0$.

Note, however, that the stability of the equilibrium is *not* the only thing that changes when p_1 passes through zero. To see this a bit more clearly, let us introduce a pair of new variables r and θ such that

$$x_1 = r \cos \theta$$
$$x_2 = r \sin \theta \qquad (8.64)$$

In that case, the derivatives $\dot{x}_1$ and $\dot{x}_2$ become

$$\dot{x}_1 = \dot{r} \cos \theta - (r \sin \theta)\dot{\theta}$$
$$\dot{x}_2 = \dot{r} \sin \theta + (r \cos \theta)\dot{\theta} \qquad (8.65)$$

Substituting (8.65) into (8.62), we can rewrite this pair of equations as

$$\dot{r} \cos \theta - (r \sin \theta)\dot{\theta} = p_1 r \cos \theta - p_2 r \sin \theta - r^3 \cos \theta$$
$$\dot{r} \sin \theta + (r \cos \theta)\dot{\theta} = p_2 r \cos \theta + p_1 r \sin \theta - r^3 \sin \theta \qquad (8.66)$$

After sorting out the terms, we obtain

$$\cos \theta (\dot{r} - p_1 r + r^3) - \sin \theta (r\dot{\theta} - p_2 r) = 0$$
$$\sin \theta (\dot{r} - p_1 r + r^3) + \cos \theta (r\dot{\theta} - p_2 r) = 0 \qquad (8.67)$$

It is not difficult to show that equation (8.67) reduces to the following two conditions:

$$\begin{array}{ll} \dot{r} - p_1 r + r^3 = 0 & \dot{r} = r(p_1 - r^2) \\ r\dot{\theta} - p_2 r = 0 & \Rightarrow \quad r(\dot{\theta} - p_2) = 0 \end{array} \qquad (8.68)$$

Both of these equations are obviously satisfied when $r = 0$ (which corresponds to the equilibrium at the origin). However, when $p_1 > 0$, there is also a solution of the form

$$r(t) = \sqrt{p_1}$$
$$\theta(t) = p_2 t \qquad (8.69)$$

In the original coordinates, this solution can be expressed as

$$x_1(t) = \sqrt{p_1} \cos p_2 t$$
$$x_2(t) = \sqrt{p_1} \sin p_2 t \qquad (8.70)$$

which is clearly a limit cycle with amplitude $\sqrt{p_1}$. A more detailed analysis shows that this limit cycle is an attractor for *all* values of $p_1 > 0$ (an illustration of this fact is provided in Fig. 8.17, for $p_1 = 1$ and $p_2 = 1$).

8.2. CATASTROPHES

One of the difficulties that we encounter in dealing with Hopf bifurcations stems from the fact that limit cycles consist of infinitely many points. As a result, we cannot directly apply the approach used in Examples 8.3-8.5, where we plotted equilibria as a function of the parameter. A simple way to resolve this problem would be to plot only the *local maxima* of a trajectory that belongs to the attractor, and add them to the equilibrium points that are already in the diagram. In Example 8.6, *all* the local maxima of the limit cycle are equal to $\sqrt{p_1}$, so its contribution for a given value of p_1 would be a single point on the bifurcation diagram. The plot for system (8.62) obtained using this method is shown in Fig. 8.18.

In concluding this section, we should note that hypersensitivity to initial conditions and parameter changes arises in discrete systems as well. As an illustration of this fact, let us consider the following simple equation

$$x_{k+1} = px_k(1 - x_k) \tag{8.71}$$

where p represents a variable parameter. For $p = 3.83$, the system exhibits completely regular behavior, whose general pattern is shown in Fig. 8.19. When this value is decreased by less than 1%, however, we see a sudden aperiodic "burst" in the midst of uniform oscillations, which clearly constitutes a qualitative change in the temporal evolution of the system (this situation is shown in Fig. 8.20).

To see how system (8.71) responds to changes in initial conditions, in Fig. 8.21 we show the difference between solutions that correspond to $x(0) = 0.5$ and $y(0) = 0.50000001$ when $p = 3.9$. Although $x(0)$ and $y(0)$ differ at the eighth decimal place, it is obvious that the two solutions aren't even remotely similar after about 50 iterations. This resembles the behavior shown in Fig. 8.10, and is a clear indicator of hypersensitivity with respect to initial conditions.

8.2 Catastrophes

Catastrophes represent a special class of bifurcations, which are characterized by an abrupt and irreversible change in the number of system equilibria. The simplest type of catastrophe is known as a *fold*, and involves a single parameter. An example of this sort would be a system of the form

$$\dot{x} = ax^2 + p \tag{8.72}$$

where $a > 0$ is a constant, and p is a slowly varying parameter. The equilibria of this system are the solutions of equation:

$$0 = ax^2 + p \tag{8.73}$$

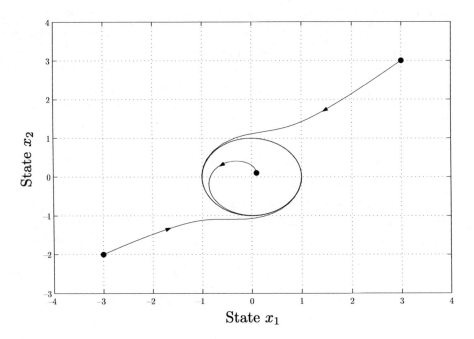

Figure 8.17: The limit cycle for a Hopf bifurcation.

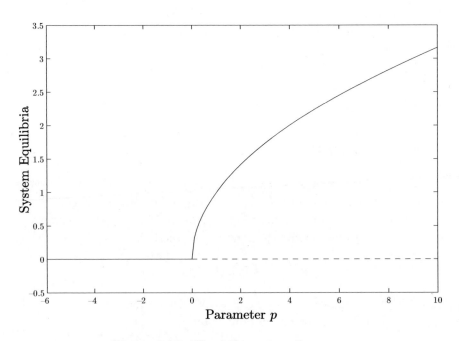

Figure 8.18: The bifurcation diagram.

8.2. CATASTROPHES

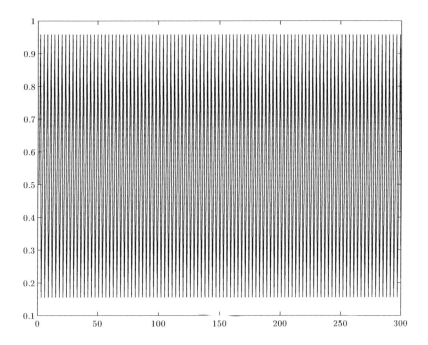

Figure 8.19: The solution of system (8.71) for $p = 3.83$.

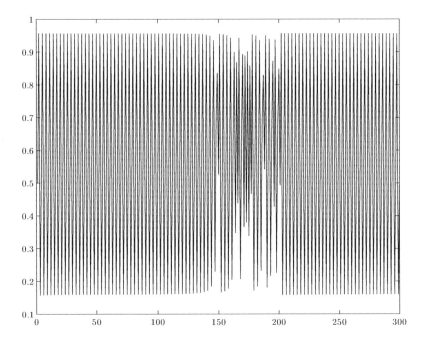

Figure 8.20: The solution of system (8.71) for $p = 3.82837$.

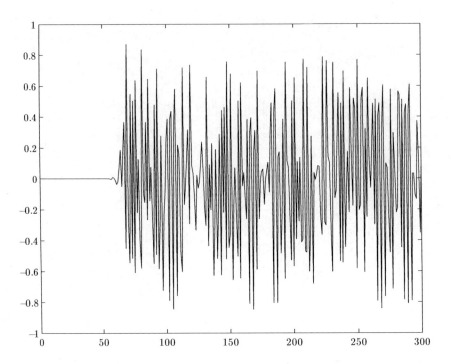

Figure 8.21: The difference $x(t) - y(t)$ for $x_0 = 0.5$ and $y_0 = 0.50000001$.

which are

$$x_1^e = \sqrt{\frac{-p}{a}} \quad \text{and} \quad x_2^e = -\sqrt{\frac{-p}{a}} \qquad (8.74)$$

These solutions obviously exist only for $p < 0$, which means that the system has no equilibrium when $p > 0$.

In order to evaluate the stability properties of these two equilibria, we need to linearize the right hand side of equation (8.72), which produces

$$f(x) \approx f'(x^e)(x - x^e) = 2ax^e(x - x^e) \qquad (8.75)$$

Setting $y = x - x^e$, we obtain

$$\dot{y} = (2ax^e)\, y \qquad (8.76)$$

which implies that x_1^e is unstable and x_2^e is stable (since $a > 0$ by assumption).

The dependence of the equilibrium on parameter p is shown in Fig. 8.22. The bifurcation occurs when $p = 0$, which is the point when both equilibria vanish. This is obviously a *qualitative* change in the system behavior, which occurs abruptly as p transitions from negative to positive values.

8.2. CATASTROPHES

The diagram in Fig. 8.22 suggests that a "fold" is not very different from a saddle node bifurcation. A far more interesting scenario arises if we consider a system of the form

$$\dot{x} = -x^3 + qx - p \tag{8.77}$$

which contains two parameters, p and q. To see how such a system behaves, let us temporarily fix q at $q = 2$, and consider what happens when p gradually changes. The equilibria of (8.77) can be found by solving equation

$$0 = -x^3 + 2x - p \tag{8.78}$$

which allows for three possible scenarios:

CASE 1. For $p < -1.089$, the equation has one real root and a pair of complex roots.

CASE 2. For $-1.089 \leq p \leq 1.089$, there are three real roots.

CASE 3. For $p > 1.089$, there is once again a single real root and two complex roots.

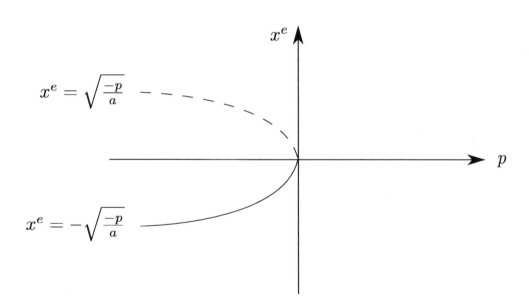

Figure 8.22: Bifurcation diagram for system (8.72).

To see which equilibria are stable and which are not, we need to linearize the right-hand side as

$$f(x) \approx f'(x^e)(x - x^e) \tag{8.79}$$

and set $y = x - x^e$. In this case, we have

$$f'(x) = -3x^2 + 2 \qquad (8.80)$$

so the linearized system becomes

$$\dot{y} = \phi(x^e)y \qquad (8.81)$$

where $\phi(x) = -3x^2 + 2$.

Function $\phi(x)$ is a parabola, whose shape is shown in Fig. 8.23. From this figure, it is not difficult to see that equilibria which lie in the interval $[-0.816 \ \ 0.816]$ will be *unstable*, since they correspond to $\phi(x) > 0$.

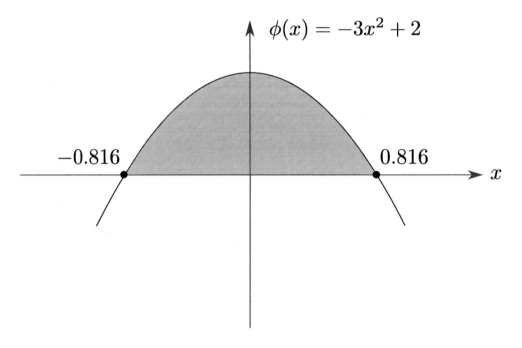

Figure 8.23: Condition for instability.

The bifurcation diagram for this system is shown in Fig. 8.24, which indicates that system (8.77) has a *single* stable equilibrium for $p < -1.089$ and $p > 1.089$. When p lies between these two values, there are two stable equilibria and one that is not (this equilibrium belongs to interval $[-0.816 \ \ 0.816]$, and is indicated by a dashed line). From this we can conclude that $p = -1.089$ and $p = 1.089$ are bifurcation points, which correspond to qualitative changes in the system.

8.2. CATASTROPHES

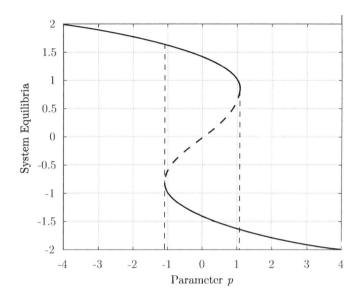

Figure 8.24: The bifurcation diagram for system (8.77) when $q = 2$.

The curve in Fig. 8.24 has a characteristic S-shape, which is a "trademark" of the so-called *cusp bifurcation*. To see why this phenomenon is referred to as a "catastrophe", let us assume that parameter p is gradually increased from -4 to 4. As we do so, each small change in p will produce a brief transient, and the system will then move to a new equilibrium (whose region of attraction includes the previous equilibrium). This is typically the equilibrium that is *closest* to the previous one, which means that the system will gradually "slide" along the top branch of the bifurcation diagram (as shown in Fig. 8.25).

It is important to note, however that there is a *discontinuous* change in the equilibrium when p crosses 1.089. Just before this value is reached, system (8.77) has three equilibria: $x_1^e = 0.82$ (stable), $x_2^e = 0.81$ (unstable) and $x_3^e = -1.62$ (stable). Since it moves along the top branch of the diagram, its solutions will converge to $x^e = 0.82$. If p is now perturbed so that its value is slightly increased, the first two equilibria disappear, and the system must switch from $x^e = 0.82$ to $x^e = -1.63$. This sudden (and very large) transition is referred to as a "catastrophe".

The effects of such a transition can be seen in Figs. 8.26 and 8.27, which show how $x(t)$ evolves for $p = 1.088$ and $p = 1.089$ (in both cases, the initial condition was chosen as $x(0) = 2$). It is evident that the two solutions look very different, even though the change in parameter p occurs at the third decimal place.

170   CHAPTER 8.   NONLINEARITY, CHAOS AND CATASTROPHES

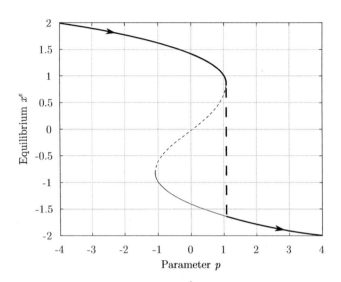

Figure 8.25: Evolution of the stable equilibrium as p varies from -4 to 4.

Something very similar happens when we vary p in the opposite direction (from $p = 4$ to $p = -4$). In this case the system behaves in the manner illustrated in Fig. 8.28, which represents a "mirror image" of the one in Fig. 8.25.

The diagram in Fig. 8.28 indicates that the equilibrium moves along in the bottom branch of the diagram as p decreases from 4 to -1. Just before p crosses $p = -1.089$ (which is the point when the catastrophe occurs), the system has three equilibria: $x_1^e = -0.82$ (stable), $x_2^e = -0.81$ (unstable) and $x_3^e = 1.62$ (stable). This means that its solutions will converge to $x_1^e = -0.82$ when parameter p is slightly perturbed. Once p decreases below -1.089, however, the system loses two of its equilibria and only $x_3^e = 1.63$ remains. As a result, there will be a discontinuous "jump" from $x^e = -0.82$ to $x^e = 1.63$.[7]

The phenomena that we have been examining so far correspond to the special case when parameter q is fixed at $q = 2$, and p is varied. In principle, we could repeat this process for a range of values of q, obtaining a different S-shaped curve each time. These diagrams could then be combined into a single three dimensional plot, which would allow us to monitor the interplay between x^e, p and q and describe their relationship in geometric terms.

A somewhat different (and arguably simpler) way to assess how parameters p and q affect the system dynamics would be to consider values of p for which the number of *stable* equilibria changes from one to two (for $q = 2$, we found that these values are $p_1 = -1.089$ and $p_2 = 1.089$). If we plot these points for

8.2. CATASTROPHES

different values of q, we obtain what is known as the *catastrophe set* (whose form is shown in Fig. 8.29).

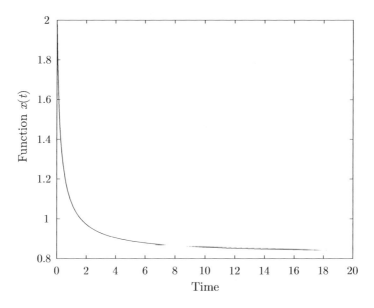

Figure 8.26: Solution $x(t)$ for $p = 1.088$.

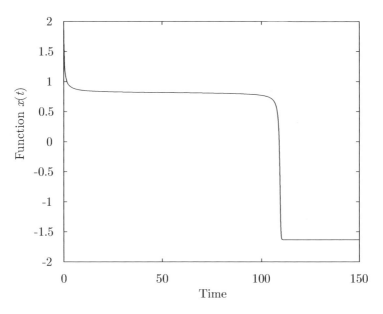

Figure 8.27: Solution $x(t)$ for $p = 1.089$.

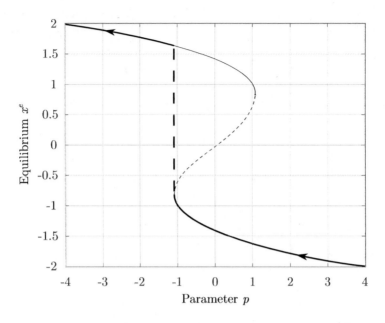

Figure 8.28: Evolution of the stable equilibrium as p varies from 4 to -4.

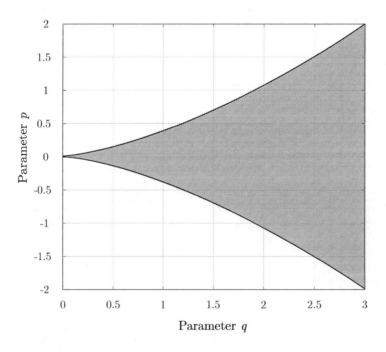

Figure 8.29: The catastrophe set for system (8.77).

8.2. CATASTROPHES

The two curves separate the parametric region with two stable equilibria from the region where there is only one. Note that a "catastrophe" occurs when we cross p_2 from below, or p_1 from above (in other words, when we *leave* the shaded area). We should also recognize that for negative values of q the system has a single stable equilibrium for all p. For such values, the third order polynomial always has a pair of complex conjugate roots.[8]

Chapter 9

Cellular Automata

9.1 Properties of Cellular Automata

A cellular automaton can be viewed as an array of linearly arranged cells such as the one shown in Fig. 9.1.

x_1	x_2	...	x_{i-1}	x_i	x_{i+1}	...	x_{n-1}	x_n

Figure 9.1: Schematic representation of a one-dimensional cellular automaton.

The dynamic behavior of such a system can be described using a truth table, which shows how the state of cell i at time $k+1$ (denoted $x_i(k+1)$) is determined by the states of cells $i-1$, i and $i+1$ at time k. Such a table has the general form shown below, where X_i can be either 0 or 1 ($i = 0, 1, \ldots, 7$).

$x_{i-1}(k)$	$x_i(k)$	$x_{i+1}(k)$	$x_i(k+1)$
0	0	0	X_0
0	0	1	X_1
0	1	0	X_2
0	1	1	X_3
1	0	0	X_4
1	0	1	X_5
1	1	0	X_6
1	1	1	X_7

Table 9.1. Description of a cellular automaton using a truth table.

It is not difficult to see that Table 9.1 represents a family of discrete mappings of the form

$$x_i(k+1) = F\left[x_{i-1}(k), x_i(k), x_{i+1}(k)\right] \tag{9.1}$$

where each choice of $\{X_0, \ldots, X_7\}$ corresponds to a *different* function F. Since there are $2^8 = 256$ such choices, we can identify each of these functions with a unique integer, which is formed as

$$N = X_7 \cdot 2^7 + X_6 \cdot 2^6 + \ldots + X_1 \cdot 2^1 + X_0 \cdot 2^0 \tag{9.2}$$

By doing so, we can easily specify which function we are referring to.

It is often convenient to represent relationship (9.2) as a cube whose vertices are labeled 0 through 7 (as shown in Fig. 9.2). If we color vertex i black when $X_i = 1$ and white when $X_i = 0$, each function F will correspond to a unique color pattern. Such a representation is sometimes preferred to the one in (9.2), because it provides us with a nice visual tool for distinguishing between different rules.

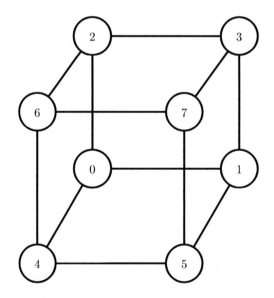

Figure 9.2: An alternative representation of the truth table.

The temporal evolution of a one-dimensional cellular automaton is typically described as a sequence of strings, which represent the states of the cells at different times. For better clarity, cells whose state is a 1 are usually colored black, while those whose state is a 0 are left blank. A typical pattern of this sort is shown in Fig. 9.3, which corresponds to the rule provided in Table 9.2

9.1. PROPERTIES OF CELLULAR AUTOMATA

(in this case, the initial configuration has a single black cell in the middle of the row).

Alternatively, we can describe the state of an automaton by converting its binary pattern into a decimal number. This approach is particularly useful when it comes to identifying limit cycles and their basins of attraction (see Figs. 9.13-9.15 for an illustration).

$x_{i-1}(k)$	$x_i(k)$	$x_{i+1}(k)$	$x_i(k+1)$
0	0	0	0
0	0	1	1
0	1	0	1
0	1	1	1
1	0	0	1
1	0	1	1
1	1	0	1
1	1	1	1

Table 9.2. Rule for a simple cellular automaton.

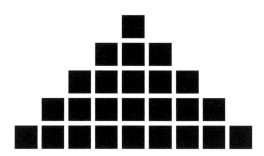

Figure 9.3: The first five steps in the evolution of the automaton described in Table 9.2.

Pattern Classes

Since each of the 256 choices for function $F(x(k))$ is quite straightforward, one would intuitively expect that the resulting patterns are simple and fairly regular. The three examples shown in Figs. 9.4-9.6 seem to confirm this conjecture.

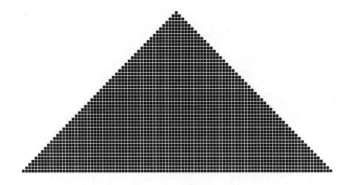

Figure 9.4: Pattern for Rule 122.

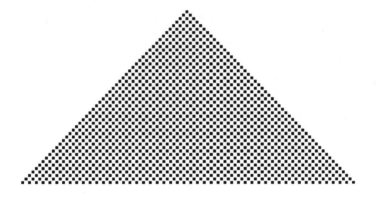

Figure 9.5: Pattern for Rule 114.

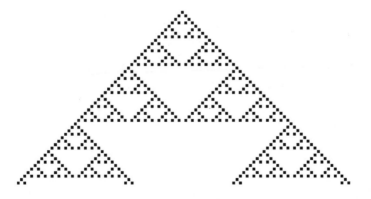

Figure 9.6: Pattern for Rule 90.

9.1. PROPERTIES OF CELLULAR AUTOMATA

Based on these figures, one would be tempted to conclude that *all* one-dimensional cellular automata exhibit some type of order in their temporal evolution (the pattern is either uniform, periodic, or has a nested, fractal-like structure). It turns out, however, that this is *not* always the case, as illustrated by the highly irregular pattern shown in Fig. 9.7. In this case, the overall structure appears to be random even after a million steps, which suggests that a set of very simple rules can sometimes produce extraordinarily complicated dynamic behavior.

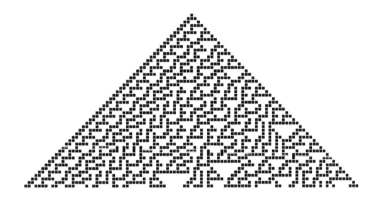

Figure 9.7: Pattern for Rule 30.

A closer examination of the 256 possible rules for one-dimensional automata indicates that there are actually four types of possible patterns, which can be classified as *uniform* (Class 1), *repetitive* (Class 2), *random* (Class 3) and *complex* (Class 4). The last of these categories corresponds to a mix of regular and irregular behavior, such as the one shown in Fig. 9.8. What we typically see in such cases are relatively simple localized structures, which can "move around" and "interact" in very complicated ways. [9]

An interesting question that arises in this context is whether the number of generic pattern types changes appreciably if we allow for a more complicated set of rules. In order to see that, let us consider a fairly simple generalization in which each cell is now allowed to have *three* different values - 0, 1 and 2. Schematically, such a situation could be represented by introducing gray as a third color, with $0 =$ white, $1 =$ gray and $2 =$ black. In this case, the triplet $[x_{i-1}(k)\ x_i(k)\ x_{i+1}(k)]$ has $3^3 = 27$ possible values, each of which can be associated with $x_i(k+1) = 0$, 1 or 2. This allows for a total of $3^{27} \approx 7.6256 \cdot 10^{12}$ different rules, which is an extraordinarily large number by any standard.

In order to simplify the analysis, in the following we will focus on a special class of rules in which the value associated with $x_i(k+1)$ depends on the *average* of $x_{i-1}(k)$, $x_i(k)$ and $x_{i+1}(k)$. Note that the sum of these three terms can take values {0, 1, 2, 3, 4, 5, 6}, which means that there are only *seven* possible averages: {0, 1/3, 2/3, 1, 4/3, 5/3, 2}. Since we can associate $x_i(k+1) = 0$, 1 or 2 with each one of them, it follows that there are $3^7 = 2,187$ different rules, which is a much more manageable number.

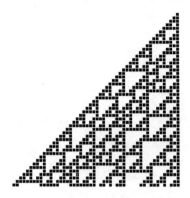

Figure 9.8: An automaton with a complex pattern.

A careful examination of the 2,187 possible scenarios shows that adding a third possible value for x_i *does not* increase the complexity of the obtained patterns in any meaningful way. This suggests that introducing more elaborate rules needn't necessarily result in a qualitative change in the overall dynamics. Indeed, it can be shown that the temporal evolution of 3-color automata can still be classified as *uniform, repetitive, random* and *complex*.[10]

Sensitivity to Initial Conditions

All the examples considered so far started from a very simple initial condition (typically, a single black cell at $t = 0$). In view of that, it would be interesting to establish whether anything changes if the initial conditions are randomized. If we compare the resulting patterns for any one of the 256 possible rules, we will indeed see certain differences (sometimes even very significant ones) when the initial conditions are altered. It turns out, however, that there are still only four basic types of behavior. The apparent "robustness" of such a classification strongly suggests that cellular automata exhibit a kind of "meta-order", which cannot be gauged from the underlying rules.

The graphs in Figs. 9.9 - 9.12 illustrate how sensitivity to initial conditions manifests itself for the four different classes of automata (in each case, only *one* cell was changed in the initial configuration). Since the last two figures are somewhat more complicated, we chose to keep the original configuration white and black, and then mark all cells that have *changed* in gray. By doing so, one

9.1. PROPERTIES OF CELLULAR AUTOMATA

can easily visualize how the perturbation in the initial conditions propagates over time.

Figure 9.9 indicates that Class 1 automata are not particularly sensitive to perturbations in the initial configuration - the effects of these disturbances eventually die out, and the system reaches the same uniform final state (which can be viewed as a sort of global attractor). Class 2 systems show a bit more variety, since changes may persist but are always localized. The behavior of Class 3 and Class 4 automata is very different in that respect, since small changes in initial conditions tend to spread throughout the system (Figs. 9.11 and 9.12 show this very clearly).

Sensitivity to changes in initial conditions is closely related to the ability of a system to process information. From Fig. 9.9, for example, we can conclude that Class 1 systems do not process information very well, since perturbations in initial conditions are "forgotten" very quickly. The situation is somewhat improved in Class 2 systems, which do retain some of the new information permanently. Fig. 9.10 shows, however, that such information is usually not communicated to other parts of the system. In contrast, Class 3 and Class 4 automata allow for long-range communication of information, since changes propagate throughout the *entire* system. This property is obviously conducive to complex dynamic behavior.

Attractors

When analyzing cellular automata, one should also keep in mind that each row has a fixed number of cells (which we will denote by L), and that this number can vary. Given that there are only 2^L possible cell configurations in such a system, its dynamic behavior is bound to become periodic after a sufficiently long time. We should note, however, that the corresponding limit cycle can sometimes be difficult to detect, since 2^L can be a very large number. When that is the case, the system dynamics appear random for all practical purposes.

Figure 9.9: The original and perturbed pattern for a Class 1 automaton.

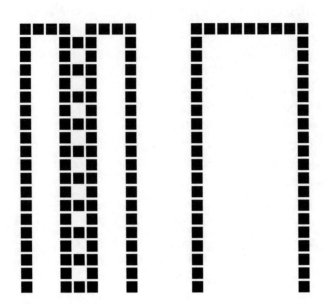

Figure 9.10: The original and perturbed pattern for a Class 2 automaton.

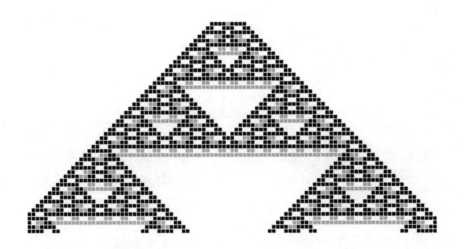

Figure 9.11: The original and perturbed pattern for a Class 3 automaton.

9.1. PROPERTIES OF CELLULAR AUTOMATA

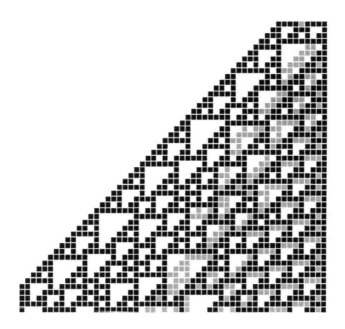

Figure 9.12: The original and perturbed pattern for a Class 4 automaton.

Depending on how L is chosen, the same rule can produce very different attractors. In general, these attractors could be both equilibria and limit cycles (although such a distinction is perhaps superfluous, since equilibria can be viewed as limit cycles of length 1). As in all other dynamic systems, each attractor has a *basin of attraction*, which includes all the initial conditions that eventually lead to it. To see what these basins look like for cellular automata, let us briefly examine Rule 62, for several different choices of L. Figure 9.13 shows that 0 is an attractor both when $L = 3$ and when $L = 4$, but the basins of this attractor are *not* the same.[11]

If we pick a larger value of L (such as $L = 9$, for example), this rule gives rise to several limit cycles of length 3, each of which acts as an attractor. The one shown in Fig. 9.14 has a large basin of attraction, while the two shown in Fig. 9.15 do not. In the latter case there actually isn't any basin to speak of - the system can enter one of these cycles *only* if its initial state is one of the states that constitute the attractor. Such attractors are sometimes referred to as "isles of Eden", since they cannot be reached from the "outside".

A particularly interesting example of an "isle of Eden" is associated with Rule 30. If the number of cells is chosen to equal $L = 27$, one of the attractors turns out to be a limit cycle of length 3,240. Leon Chua reports that this is the longest known period for an "isle of Eden", which makes Rule 30 an intriguing case study.[12] Part of the challenge in analyzing this system is to show that the limit cycle cannot be reached from the "outside", which entails checking all 2^{27} possible initial configurations.

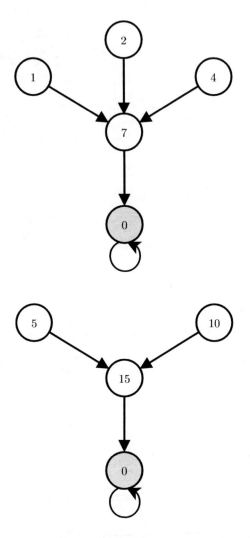

Figure 9.13: Basins of attraction for state 0 with $L = 3$ and $L = 4$.

Reversibility and Entropy

The rules that define a cellular automaton allow us to uniquely determine $x_i(k+1)$ if $x_i(k)$, $x_{i-1}(k)$, $x_{i+1}(k)$ and the initial state of the system are known. But is it possible to do the opposite, and determine the initial state given the rules and *final* configuration? In most cases, the answer to this question in *no*.

9.1. PROPERTIES OF CELLULAR AUTOMATA

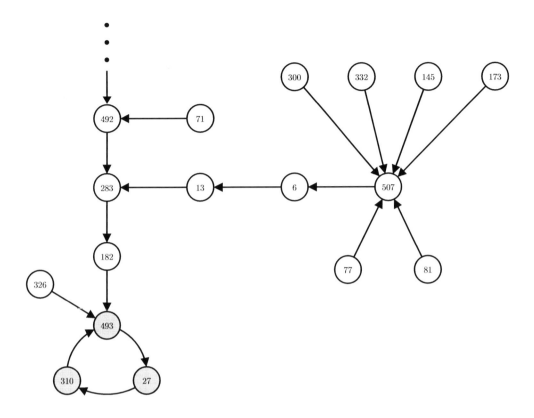

Figure 9.14: Basin of attraction for a limit cycle that corresponds to $L = 9$.

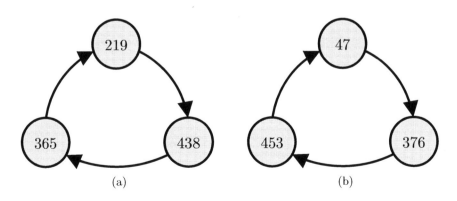

Figure 9.15: Two "Isles of Eden" for Rule 62 with $L = 9$.

To see why this is so, it suffices to consider the simple rule shown in Table 9.3 (so-called Rule 0). If we are told that the final configuration of this system is a row of zeros, it is impossible to determine what the row above it looks like, since in this case $x_i(k+1) = 0$ corresponds to as many as 8 different combinations of $x_i(k)$, $x_{i-1}(k)$, $x_{i+1}(k)$. As a result, we have no way to uniquely identify the initial state of the system from the available information.

$x_{i-1}(k)$	$x_i(k)$	$x_{i+1}(k)$	$x_i(k+1)$
0	0	0	0
0	0	1	0
0	1	0	0
0	1	1	0
1	0	0	0
1	0	1	0
1	1	0	0
1	1	1	0

Table 9.3. Example of an irreversible automaton.

Let us now contrast this example to the rule shown in Table 9.4. It is not difficult to see that in this case $x_i(k+1)$ is simply the opposite of $x_i(k)$ (the other two states have no influence whatsoever). Under such circumstances, it is obviously possible to backtrack, and recover row k from the one below it. We can therefore start from the final configuration and use the rule to reconstruct the initial condition (whenever this is possible, we say that the system is *reversible*) [13].

$x_{i-1}(k)$	$x_i(k)$	$x_{i+1}(k)$	$x_i(k+1)$
0	0	0	1
0	0	1	1
0	1	0	0
0	1	1	0
1	0	0	1
1	0	1	1
1	1	0	0
1	1	1	0

Table 9.4. Example of a reversible automaton.

In order to gain a better understanding of what reversibility means in practice, it is helpful to examine a class of cellular automata where this property is

9.1. PROPERTIES OF CELLULAR AUTOMATA

guaranteed by construction. In systems of this sort, it is assumed that $x_i(k+1)$ depends *both* on $x_i(k-1)$ *and* on vector $[x_i(k)\ x_{i-1}(k)\ x_{i+1}(k)]$. A schematic representation of one such rule is shown in Fig. 9.16, which has $2^4 = 16$ different input configurations.

Rules such as the one shown in Fig. 9.16 are *symmetric*, in the sense that they can be used to determine both future and past configurations. Figure 9.17 shows what happens when we exploit this feature, and run the system "in reverse" starting from the two final states. It is not difficult to see that this procedure allows us to fully recover the original initial condition (which, in this case, is a single black cell in the middle of the row).

We should note, however, that something like this is virtually impossible to replicate in experiments with real physical systems, since we usually have no way of reconstructing their "history" from their present state. This observation brings up an interesting question about the relationship between natural laws (as we understand them, of course) and their physical manifestations. On the surface this relationship appears to be somewhat paradoxical, since our mathematical models tend to be symmetric with respect to time, but the processes and phenomena that they describe rarely exhibit any form of reversibility. Almost all of them evolve in a way that corresponds to an increase in entropy (or equivalently, an increase in disorder), which allows us to clearly distinguish between past, present and future states.

What we have learned about cellular automata in recent years suggests a possible explanation for this discrepancy, which is based on the recognition that irreversible outcomes are extremely likely, but *not* imminent. To get a sense for what this means, we should observe that reversible cellular automata often exhibit behavior that appears to be completely random (and therefore irreversible for all practical purposes). The system in Fig. 9.18(b) provides a perfect illustration of this property, since it starts out with a regular pattern, but eventually loses all discernible structure.

What is particularly interesting about this rule is the fact that we can run it *in reverse* as well (in which case our previous final configuration would become the *initial* one). If we do so, we will ultimately obtain the pattern shown in Fig. 9.18(c). This figure tells us that under certain very special initial conditions, an initially irregular configuration can eventually evolve into one that exhibits elements of order.

To put this result in the proper perspective, we should note that the behavior of such systems is extraordinarily sensitive to initial conditions. Indeed, the patterns shown in Figs. 9.18(a) and 9.18(d) suggest that most initial configurations produce either completely random or very orderly behavior. Only rarely do we encounter situations such as the one in Fig. 9.18(c), which are characterized by an increase in order over time.

If we were to extend this line of reasoning to physical systems, we could argue that certain special initial configurations could potentially lead to a *decrease* in entropy. The reason why we never observe such behavior in nature is not because it is impossible, but because the conditions that give rise to it are highly improbable.

Figure 9.16: A reversible cellular automaton.

Figure 9.17: The pattern when the automaton in Fig. 9.16 is run "in reverse".

9.1. PROPERTIES OF CELLULAR AUTOMATA 189

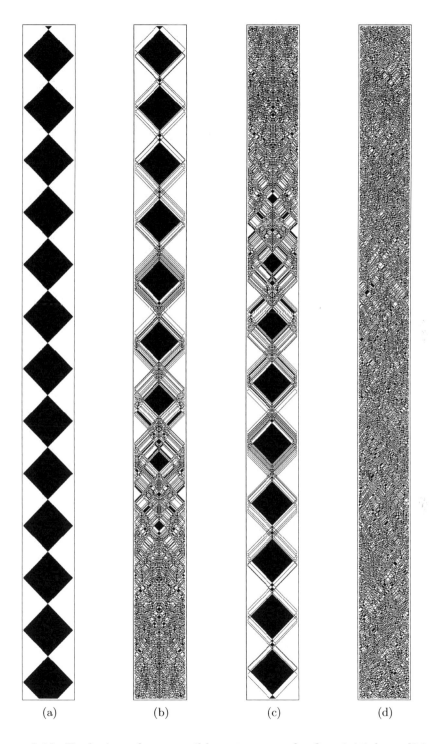

Figure 9.18: Evolution of a reversible automaton for four initial conditions.

9.2 An Analytic Classification of Rules

Mathematician Stephen Wolfram was the first to systematically study cellular automata, and recognize that they can produce a wide variety of dynamic patterns, ranging from simple to highly complex. Based on the characteristics of these patterns, he grouped the 256 possible rules into the four different classes described in Section 9.1. Wolfram was aware, of course, that such a classification is purely empirical, but he argued that there appears to be no alternative, since cellular automata can only be studied experimentally.

Some recent results obtained by Leon Chua challenge such a conclusion, and allow us to examine certain important properties of cellular automata using the tools of nonlinear system theory.[14] Although these results fail to predict exactly how a given cellular automaton will evolve over time, they do provide us with a set of analytical criteria for classifying the different forms of dynamic behavior that are observed. In this section we will take a closer look at these criteria, and examine their practical implications.

Chua's approach is based on a hypothetical dynamic system whose schematic description is provided in Fig. 9.19. The three inputs are *discrete*, and correspond to $x_{i-1}(k)$, $x_i(k)$ and $x_{i+1}(k)$, respectively. These quantities are not strictly Boolean, however, since they take values -1 and 1 instead of 0 and 1.

Unlike the inputs, the internal state $z_i(t)$ is assumed to be a *continuous* function of time, as is the output $y_i(t)$. The equations that describe how these two variables evolve over time have the general form

$$\dot{z}_i(t) = f(z_i(t), x_{i-1}(k), x_i(k), x_{i+1}(k))$$
$$y_i(t) = \varphi(z_i(t))$$
(9.3)

where functions f and φ are chosen so that the system has at least one stable equilibrium that can be reached from initial condition $z_i(0)$. The value of the output that corresponds to this equilibrium is denoted y_i^e, and is assumed to equal $x_i(k+1)$.

In the model proposed by Chua, functions f and φ are defined as [15]

$$f(z_i(t), x_{i-1}(k), x_i(k), x_{i+1}(k)) = -z_i + |z_i+1| - |z_i-1| + w(x_{i-1}, x_i, x_{i+1})$$ (9.4)

and

$$\varphi(z_i(t)) = 0.5(|z_i + 1| - |z_i - 1|)$$ (9.5)

The term $w(x_{i-1}, x_i, x_{i+1})$ in (9.4) has the form

$$w(x_{i-1}, x_i, x_{i+1}) = \rho_2 + c_1|[\rho_1 + c_2|\rho_0 + b_1 x_{i-1} + b_2 x_i + b_3 x_{i+1}|]|$$ (9.6)

where b_1, b_2, b_3, ρ_0, ρ_1, ρ_2 are real numbers, while c_1 and c_2 can only take values -1 or 1. Given that there are no restrictions on the first six parameters,

9.2. AN ANALYTIC CLASSIFICATION OF RULES

the number of functions that can be constructed in this manner is obviously unlimited.

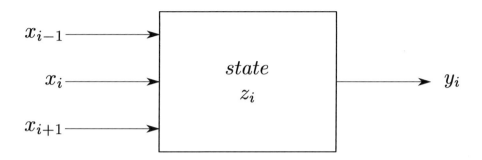

Figure 9.19: Chua's hypothetical dynamic system.

If we define an auxiliary variable

$$\sigma = b_1 x_{i-1} + b_2 x_i + b_3 x_{i+1} \tag{9.7}$$

the function in (9.6) can be expressed in a more compact form as

$$w(\sigma) = \rho_2 + c_1 |[\rho_1 + c_2|\rho_0 + \sigma|]| \tag{9.8}$$

The overall system (9.3) can then be rewritten as

$$\begin{aligned} \dot{z}_i &= -z_i + |z_i + 1| - |z_i - 1| + w(\sigma) \\ y_i &= 0.5(|z_i + 1| - |z_i - 1|) \end{aligned} \tag{9.9}$$

The following theorem highlights an important property of any such system, which is independent of the way in which the eight parameters that define $w(\sigma)$ are chosen.

Theorem 9.1. When $z_i(0) = 0$, any system of the form (9.9) will necessarily converge to a stable equilibrium z_i^e. This equilibrium satisfies $z_i^e > 2$ when $w(\sigma) > 0$, and $z_i^e < -2$ when $w(\sigma) < 0$.

Proof. A graph of the function

$$g(z_i) = -z_i + |z_i + 1| - |z_i - 1| \tag{9.10}$$

which appears on the right hand side of equation (9.9) is shown in Fig. 9.20. Depending on whether $w(\sigma)$ is positive or negative, we have two possible scenarios, which are illustrated in Figs. 9.21 and 9.22. Note that in both cases, the system trajectory starts on the vertical axis, since we assumed that $z_i(0) = 0$.

In the graph in Fig. 9.21 (which corresponds to $\dot{z}_i(0) = w(\sigma) > 0$) the trajectory necessarily moves to the *right*, since $\dot{z}_i(t) > 0$ implies that $z_i(t)$ must *increase* over time. The point labeled $z_i^e(Q_+)$ represents an equilibrium of the system, since it satisfies

$$\dot{z}_i = g(z_i) + w(\sigma) = 0 \qquad (9.11)$$

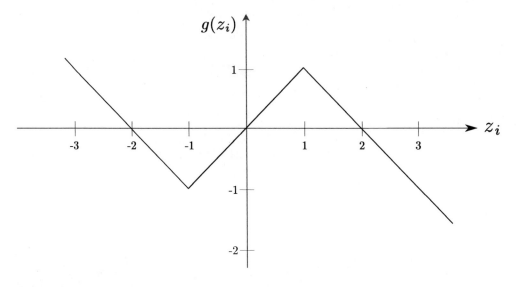

Figure 9.20: Function $g(z_i)$.

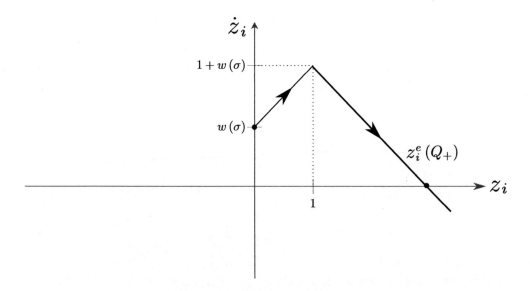

Figure 9.21: The phase plot in case $w(\sigma) > 0$.

9.2. AN ANALYTIC CLASSIFICATION OF RULES

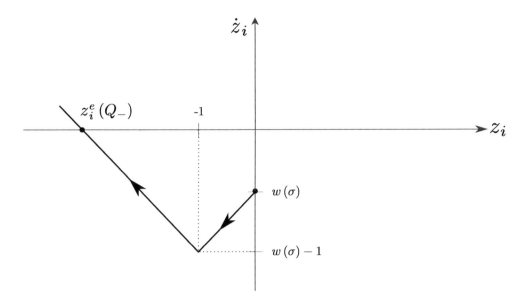

Figure 9.22: The phase plot in case $w(\sigma) < 0$.

The specific value of $z_i^e(Q_+)$ depends on $w(\sigma)$ (and therefore on inputs $x_{i-1}(k)$, $x_i(k)$ and $x_{i+1}(k)$), but it is easily verified that this value is always greater than 2, since $g(z_i)$ becomes negative for $z_i > 2$. Equilibrium $z_i^e(Q_+)$ is an attractor, since $\dot{z}_i$ becomes *negative* as we move to the right of this point. Any such disturbance would make $z_i(t)$ *decrease*, "pushing" the trajectory to the left and eventually back to $z_i^e(Q_+)$.[16]

The scenario shown in Fig. 9.22 is similar to the one analyzed above, except for the fact that $w(\sigma)$ is now negative. This means that $\dot{z}_i(t) < 0$, so the trajectory will move to the *left* and eventually converge to equilibrium $z_i^e(Q_-)$. This equilibrium is an attractor, since any disturbance that would move the system further to the left would make $z_i(t)$ *increase* (because $\dot{z}_i$ becomes *positive*). It must also satisfy $z_i^e(Q_-) < -2$, since $g(z_i)$ becomes positive for $z_i < -2$.
Q.E.D.

Chua showed that for *any* given cellular automaton, the eight parameters in expression (9.6) can be chosen so that the steady state values y_i^e match the desired $x_i(k+1)$. The proof of this property is not particularly elegant, since it amounts to empirically identifying an appropriate set of parameters for each of the 256 possible rules. Nevertheless, it is a perfectly valid approach, despite its reliance on a "trial and error" procedure.

To get a sense for how this proof works, let us consider the discrete function described in Table 9.5, in which all occurrences of 0 have been replaced with -1. This function is commonly referred to as Rule 110, since the "Boolean equivalent" of the last column in the table corresponds to

$$N = 0\cdot 2^7 + 1\cdot 2^6 + 1\cdot 2^5 + 0\cdot 2^4 + 1\cdot 2^3 + 1\cdot 2^2 + 1\cdot 2^1 + 0\cdot 2^0 = 110 \quad (9.12)$$

This rule is particularly interesting in the study of cellular automata, since it gives rise to some very complex and intricate patterns.

$x_{i-1}(k)$	$x_i(k)$	$x_{i+1}(k)$	$x_i(k+1)$
−1	−1	−1	−1
−1	−1	1	1
−1	1	−1	1
−1	1	1	1
1	−1	−1	−1
1	−1	1	1
1	1	−1	1
1	1	1	−1

Table 9.5. Rule 110 where all zeros are replaced by −1.

If we choose $b_1 = 1$, $b_2 = 2$, $b_3 = -3$, $\rho_0 = -1$, $\rho_1 = 0$ and $\rho_2 = -2$, and set $c_1 = c_2 = 1$, we obtain

$$\sigma = x_{i-1} + 2x_i - 3x_{i+1} \quad (9.13)$$

and

$$w(\sigma) = -2 + |\sigma - 1| \quad (9.14)$$

Since $z_i(0)$ is assumed to be zero, we know by virtue of Theorem 9.1 that a system of the form (9.9) always converges to a stable equilibrium z_i^e, which is greater than 1 when $w(\sigma) > 0$ and less than −1 when $w(\sigma) < 0$. In the first case, we have that

$$y_i^e = 0.5(|z_i^e + 1| - |z_i^e - 1|) = 0.5[(z_i^e + 1) - (z_i^e - 1)] = 1 \quad (9.15)$$

and in the second

$$y_i^e = 0.5(|z_i^e + 1| - |z_i^e - 1|) = 0.5[-(z_i^e + 1) + (z_i^e - 1)] = -1 \quad (9.16)$$

so we can conclude that

$$y_i^e = \begin{cases} 1, & \text{if } w(\sigma) > 0 \\ -1, & \text{if } w(\sigma) < 0 \end{cases} \quad (9.17)$$

or equivalently

$$y_i^e = \text{sgn}\{w(\sigma)\} \quad (9.18)$$

9.2. AN ANALYTIC CLASSIFICATION OF RULES

This simple relationship allows us to compute y_i^e for each of the 8 possible input combinations. The results obtained for Rule 110 are shown in Table 9.6, which also includes the corresponding values of σ and $w(\sigma)$. It is easily verified that the last column in Table 9.6 is identical to the last column in Table 9.5, which means that the steady state output of dynamic system (9.9) emulates this rule.[17]

$x_{i-1}(k)$	$x_i(k)$	$x_{i+1}(k)$	σ	$w(\sigma)$	y_i^e
-1	-1	-1	0	-1	-1
-1	-1	1	-6	5	1
-1	1	-1	4	1	1
-1	1	1	-2	1	1
1	-1	-1	2	-1	-1
1	-1	1	-4	3	1
1	1	-1	6	3	1
1	1	1	0	-1	-1

Table 9.6. The values of σ, $w(\sigma)$ and y_i^e for rule 110.

The model proposed in (9.9) allows us to classify the 256 possible automata in a way that is *not* purely empirical, and relies on precise mathematical criteria. To see how this may be done, let us once again consider Rule 110, and examine the corresponding function $w(\sigma)$ a bit more closely. In this case, $w(\sigma)$ has the form shown in (9.14), which can be equivalently represented as

$$w(\sigma) = \begin{cases} \sigma - 3, & \text{if } \sigma \geq 1 \\ -\sigma - 1, & \text{if } \sigma \leq 1 \end{cases} \tag{9.19}$$

A graph of this function is provided in Fig. 9.23, in which the two transition points where $w(\sigma)$ changes sign are labeled T_1 and T_2, respectively. The plot also displays the values of σ that correspond to the 8 different combinations of inputs x_{i-1}, x_i and x_{i+1}. These values are marked by solid black dots on the σ-axis, and are labeled according to their binary representation.

Recalling equation (9.18), it follows that the graph in Fig. 9.23 effectively partitions the input combinations into three groups. The one in the middle corresponds to $y_i^e = -1$ (since $w(\sigma) < 0$), while the groups to the left and right of it correspond to $y_i^e = 1$ (since $w(\sigma) > 0$). Note that the first group would be represented by white nodes in the "cube representation" that was introduced in Fig. 9.2, and the other two by black ones.

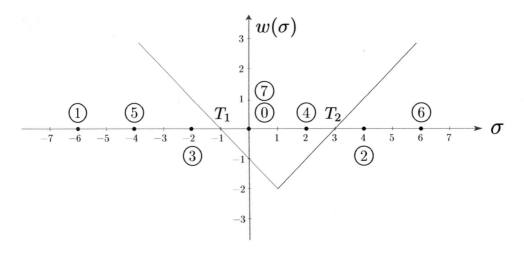

Figure 9.23: Function $w(\sigma)$ and the corresponding partitioning of input combinations.

In order to interpret what this partitioning really means, we should observe that the two transition points satisfy equation

$$w(\sigma) = -2 + |\sigma - 1| = -2 + |x_{i-1} + 2x_i - 3x_{i+1} - 1| = 0 \tag{9.20}$$

when the parameters are chosen to emulate Rule 110. For the transition point that corresponds to $\sigma = 3$ this equation becomes

$$w(\sigma) = \sigma - 3 = x_{i-1} + 2x_i - 3x_{i+1} - 3 = 0 \tag{9.21}$$

while for $\sigma = -1$ we have

$$w(\sigma) = -\sigma - 1 = -(x_{i-1} + 2x_i - 3x_{i+1}) - 1 = 0 \tag{9.22}$$

These equations describe two different planes in the three dimensional "x-space", which effectively separate the black nodes from the white ones in the cube.

Although each of the 256 possible rules allow for this sort of partitioning, they differ in the minimal number of planes that are required. Certain rules (such as Rule 232, for example) require only a *single* plane, while others (like Rule 150) require as many as *three* planes. The minimal number of planes needed to separate black nodes from white ones can be uniquely determined for each rule. This number (denoted in the following by κ) tells us how complex the pattern generated by the rule will be, which is why Chua referred to it as the *complexity index*.

Rules that are characterized by $\kappa = 1$ are said to be *linearly separable*, and exhibit only the simplest forms of dynamic behavior. There are 104 such rules, while the remaining 152 (which have indices $\kappa = 2$ or $\kappa = 3$) are *linearly non-separable*. This latter group of automata can produce highly complex patterns, so one can reasonably conclude that $\kappa = 2$ represents a "threshold of complexity" for elementary cellular automata.

It is important to reiterate that the complexity index allows us to classify the 256 rules based solely on their *mathematical description*, with no need for simulation. This approach is very different from the one used by Wolfram, who identified his four classes of rules by observing how they evolve over time. For the most part, these two classifications are consistent with each other, but there are some exceptions. These exceptions suggest that experimental observations can sometimes be deceptive, and can fail to detect certain subtle regularities in the system.

9.3 Random Boolean Networks

Random Boolean networks represent a generalization of cellular automata in which each cell is allowed to evolve according to a *different* rule of the form

$$x_i(k+1) = F_i(x(k)) \qquad (9.23)$$

The term $x(k)$ in (9.23) denotes the entire vector $x(k) = [x_1(k) \ x_2(k) \ \ldots \ x_n(k)]^T$, which means that $x_i(k+1)$ could depend on any number of other cells (as opposed to cellular automata, which are limited to cells $i-1$, i and $i+1$).

Boolean networks were introduced by theoretical biologist Stuart Kauffman, who used them to model the dynamic behavior of genetic systems.[18] The question that Kauffman was most interested in can be paraphrased as follows: Given that the human genome consists of roughly 100,000 genes (each of which can be "on" or "off"), why is it that only a miniscule subset of the $2^{100,000}$ possible combinations is actually realized in our bodies? In searching for an answer, Kauffman made a number of important discoveries which we will briefly survey in this section.

We begin by observing that Boolean networks can be represented as directed graphs whose nodes are assigned dynamically varying states. In such a graph, the incoming edges associated with node i indicate which "neighboring" nodes influence its state $x_i(k)$ (which can be 0 or 1 in any given step).

To get a better idea of how this works, let us consider the simple Boolean network shown in Fig. 9.24, which consists of 3 nodes. We will assume that the associated functions $F_i(x(k))$ are defined in the manner indicated in Tables 9.7-9.9. Using these three definitions, we can easily describe how the overall state vector $x(k) = [x_1(k) \ x_2(k) \ x_3(k)]^T$ evolves over time (Table 9.10 shows how this can be done).[19]

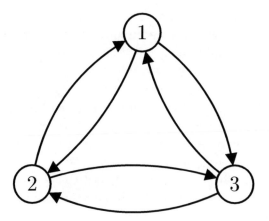

Figure 9.24: A Boolean network with $N = 3$ and $K = 2$.

Since the network in Fig. 9.24 has a total of $2^3 = 8$ possible states, we know that the length of its limit cycles cannot exceed $L = 8$. This makes the number of different configurations manageable, and allows us to identify all of them precisely. Note, however, that such a task would become virtually impossible if we were to consider a network with, say, $N = 1,000$ nodes, since the largest limit cycle could now have as many as $2^{1,000}$ steps. Cycles of that length would be undetectable for all practical purposes, and the system dynamics would seem completely random to an observer who did not know how the sequence was generated.

Because Kauffman was interested in using Boolean networks to model genetic systems he needed them to meet three important requirements, each of which represents an essential feature of living organisms.

Requirement 1. The network had to have a relatively small number of attractive limit cycles, whose length is short (even in cases when the number of nodes is very large). Under such circumstances, the different limit cycles could be identified with different cell types, and their length could potentially be associated with the duration of the cell cycles.

Requirement 2. The attractive limit cycles in the network needed to have relatively large basins of attraction. This would ensure that small perturbations (such as flipping a single bit, for example) have little or no effect on the overall operation of the system. If such a change were to occur while the system is going through its regular cycle, the disruption would be temporary and fairly minimal.

Requirement 3. The network would need to be "resilient" with respect to changes in functions $F_i(x(k))$, which would allow us to treat such disruptions as "mutations".

9.3. RANDOM BOOLEAN NETWORKS

To see whether Boolean networks can satisfy these requirements, let us consider the eight possible initial conditions in our 3-node example. In this case, it turns out that there are only three limit cycles.

$x_2(k)$	$x_3(k)$	$F_1\,[\,x_2(k), x_3(k)]$
0	0	0
0	1	0
1	0	0
1	1	1

Table 9.7. Function $F_1\,[\,x(k)]$ which defines $x_1(k+1)$.

$x_1(k)$	$x_3(k)$	$F_2\,[\,x_1(k), x_3(k)]$
0	0	0
0	1	1
1	0	1
1	1	1

Table 9.8. Function $F_2\,[\,x(k)]$ which defines $x_2(k+1)$.

$x_1(k)$	$x_2(k)$	$F_3\,[\,x_1(k), x_2(k)]$
0	0	0
0	1	1
1	0	1
1	1	1

Table 9.9. Function $F_3\,[\,x(k)]$ which defines $x_3(k+1)$.

$x_1(k)$	$x_2(k)$	$x_3(k)$	$x_1(k+1)$	$x_2(k+1)$	$x_3(k+1)$
0	0	0	0	0	0
0	0	1	0	1	0
0	1	0	0	0	1
0	1	1	1	1	1
1	0	0	0	1	1
1	0	1	0	1	1
1	1	0	0	1	1
1	1	1	1	1	1

Table 9.10. Dependence of $x(k+1)$ on $x(k)$.

(1) If the system starts from $x(0) = [0\ 0\ 0]$, it will remain in this state indefinitely, and the limit cycle will have length $L = 1$.

(2) If the system starts from $x(0) = [0\ 0\ 1]$ or $x(0) = [0\ 1\ 0]$ it will oscillate between these two states, which corresponds to a limit cycle of length $L = 2$.

(3) If the system starts from *any* other $x(0)$, it will eventually end up in state $x^* = [1\ 1\ 1]$, which is a limit cycle of length $L = 1$.

The three scenarios outlined above are schematically described in Fig. 9.25. This diagram indicates that the first two limit cycles attract only solutions that originate *in the cycle itself*. Only the third one shows genuine attractive properties, and has a basin of attraction that consists of four states (which conforms to Requirement 1).

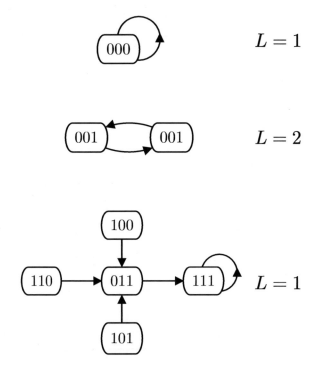

Figure 9.25: Possible limit cycles and their basins of attraction.

To see how this system fares with respect to the second requirement, let us take a closer look at state $x^* = [1\ 1\ 1]$, which is an attractive limit cycle of length $L = 1$. If we flip a single bit in x^*, the system will move into one of the following three states (which represent possible new initial conditions for the system).

a) $x(0) = [0\ 1\ 1]$

b) $x(0) = [1\ 0\ 1]$

c) $x(0) = [1\ 1\ 0]$

It is easily recognized that all three initial conditions lie in the basin of attraction of state $x^* = [1\ 1\ 1]$, so the system will return to this point after at most two steps. Contrast that to $x^* = [0\ 0\ 0]$, where every one-bit perturbation leads to a different attractor - $x(0) = [1\ 0\ 0]$ leads to $x^* = [1\ 1\ 1]$, while $x(0) = [0\ 1\ 0]$ and $x(0) = [0\ 0\ 1]$ give rise to a limit cycle of length $L = 2$.

To see whether the third requirement is realistic, let us consider a "mutation" of function $F_2[x(k)]$, which now takes the form shown in Table 9.11 (we will assume that functions F_1 and F_3 remain unchanged).

$x_1(k)$	$x_3(k)$	$F_2[x_1(k), x_3(k)]$
0	0	0
0	1	0
1	0	0
1	1	1

Table 9.11. Function $F_2[x(k)]$ after a "mutation".

The schematic diagram in Fig. 9.26 indicates that we will once again have three limit cycles, but they will be different ones (state $[0\ 0\ 0]$ is now an attractor with a sizable basin, while $x^* = [1\ 1\ 1]$ is vulnerable to any kind of perturbation). We can therefore conclude that the "mutation" does not affect the system significantly - although the specific cycles are now different, the number of "attractors" and their average length is the same as before (as is the size of the basin). This implies that the key "meta-properties" of the system are preserved.

Properties of Large Boolean Networks

In order to evaluate whether these requirements hold in larger networks, Kauffman focused his attention on systems with N nodes, each of which had exactly K neighbors. Although this simplified the problem to some extent, the number of possibilities was still enormous, since there are many different ways to construct a network with a given N and K, and just as many ways to choose functions $F_i(x(k))$. He therefore concluded that the only reasonable way to proceed was to examine *randomly sampled NK networks*, and see whether any general patterns could be identified.

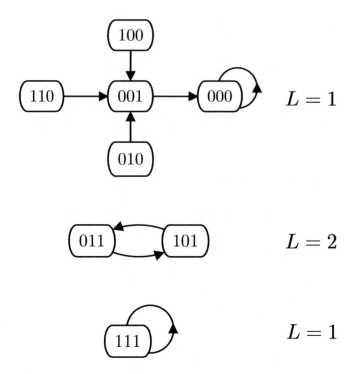

Figure 9.26: Limit cycles that correspond to the new function $F_2[x(k)]$.

In order to do that, Kauffman formed a directed graph by randomly picking K incoming edges for each node. He then constructed functions $F_i(x(k))$ by flipping a coin for every possible combination of input arguments $[x_1(k)\ x_2(k)\ \ldots\ x_N(k)]$ (the outcome of the toss would decide whether the corresponding state $x_i(k+1)$ should be 0 or 1). Once the network was formed, Kauffman simulated its evolution starting from a range of different initial conditions. He then recorded the overall number and length of the attractive limit cycles that he observed, as well as their sensitivity to perturbations and "mutations" in functions F_i.

Such experiments were performed repeatedly, each time with a different connection pattern and a different set of functions F_i (the parameters N and K, however, were kept constant). From the acquired data, Kauffman was able to estimate the average number and length of attractive limit cycles for any given family of NK networks. For $K = 1$, he found that the network quickly falls into a very short limit cycle in which most $x_i(k)$ are "frozen" at 1 or 0. Although this sort of behavior nominally satisfies Requirements 1-3, Kauffman came to the conclusion that it was too "static" to be a viable model for actual biological processes. On the other extreme (when $K = N$) he observed that the network dynamics had all the characteristics of a random process, with limit cycles whose average length increased exponentially with N. This clearly violated Requirement 1, which calls for a small number of short limit cycles.

Networks with $K = N$ also violated Requirements 2 and 3, since they exhibited extreme sensitivity to perturbations. Kauffman found that in most cases changing even a single bit would move the system to a different attractor. Since there were many attractors to choose from (each of which typically corresponded to a very long limit cycle) such a small disturbance would almost certainly result in very different dynamic behavior. Something similar happened when a single function F_i was perturbed - typically, all the cycles would change, as would their basins of attraction.

The most interesting results were observed for $K = 2$. In this case, Kauffman found that both the average length and the number of attractive limit cycles was proportional to $\sqrt{N}$. For a network with $N = 100,000$, this meant that the system quickly settled into one of roughly 300 cycles, none of which contained more than a few hundred states. Kauffman also established that such networks are quite robust with respect to random perturbations in initial conditions and functions F_i. In general, such disturbances produced only modest changes in the attractors and their basins. Even more importantly, he showed that networks with $K = 2$ exhibit a certain amount of order, although there was no "master plan" that governs the system behavior. This order arose spontaneously, and was simply the result of *local interactions*.

To put these results in the proper perspective, we should recall that one of Kauffman's central assumptions was that each attractor in a genetic network corresponds to a different *cell type*. Since the human genome has $N \approx 100,000$ genes and 256 different cell types, $\sqrt{N} = 317$ appeared to be a pretty good approximation. This led Kauffman to hypothesize that the process of cell differentiation in humans could be modeled using a Boolean network with $K = 2$. What made this conjecture particularly interesting was the fact that the "$\sqrt{N}$ law" is actually common to many organisms, from the simplest bacteria to mammals. It not only relates the number of genes to the number of cell types, but also provides a good approximation for the length of the cell cycle.

In evaluating Kauffman's model, it is also important to keep in mind that cell types must be locally stable with respect to perturbations. This property (which is known as *homeostasis*) is essential to life, since it ensures that the normal cell cycle will be restored following minor disruptions. We should note, however, that this sort of stability cannot be absolute, and must have certain limits. If that were not the case, it would be impossible to "push" the system into a new basin of attraction (which is occasionally necessary in the process of cell differentiation).

To what extent do NK networks exhibit this sort of behavior? In order to answer that question, we should first observe that the spatial appearance of limit cycles in the network depends significantly on how parameter K is chosen.

In a system with $K = 1$, for example, the overwhelming majority of nodes in the typical cycle is "frozen" at 0 or 1, with a few "islands" where the states of the nodes change. Since these "islands" account for the entire temporal variation of $x(k)$, it is fair to say that the limit cycles in such networks consists of many "fixed" nodes with a few "clusters of activity". It is also important to keep in mind that these islands are isolated both spatially and functionally, which means that a perturbation such as bit-flipping or a "mutation" in F_i would affect only the island in which it originated. There is, in other words, no mechanism through which such changes could propagate globally.

In the case when $K = N$ we have the opposite situation, where most of the nodes are constantly changing their states and only a few are "frozen". Under such circumstances, a small perturbation could easily unleash a cascade of uncoordinated events in the network, which is a trademark of "chaotic" dynamics. This is clearly not what one would expect of living cells, whose operation needs to be reasonably stable over a broad range of different conditions.

For $K = 2$ (which is the network configuration that Kauffman associated with living organisms), we have an interesting balance between the two extreme scenarios considered above. In this case, groups of "active nodes" that used to be isolated suddenly become connected into a large sub-network. Although much of the overall network remains frozen, local perturbations in any given island can now potentially propagate throughout the entire system (including distant locations). These are precisely the conditions that are needed for homeostasis - possible disturbances do not produce massive and uncontrolled changes in the network, but there is enough flexibility to allow the system to occasionally move into the basin of a different limit cycle.

Chapter 10

Self-Organized Criticality

10.1 Sand Piles and Autocatalysis

How do physical systems evolve into a state where complex dynamic behavior becomes possible, and why do they remain in such a state for prolonged periods of time? One of the necessary conditions for this is for the system to be *open*, which means that it can exchange matter and energy with the external environment. When this is the case, the constant inflow of energy allows it to occupy a state that is "far from equilibrium", in which novel forms of organization can spontaneously arise.

In the following section we will take a closer look at two systems that have the ability to self-organize (albeit in very different ways). What these two systems have in common is the fact that their basic "ingredients" are continuously supplied from the "outside", and that their behavior changes abruptly when the system reaches a critical threshold.

10.1.1 Sand Piles

Sand piles provide a simple and very helpful model for understanding how a system can self-organize, and what is meant by "criticality". What is interesting about this model is that it captures all the essential features of complex behavior, although its dynamics are described by only a few elementary rules. These rules can take many different forms, but they all assume that the sand pile is located on a flat surface, and that individual grains are added one at a time (possibly at different locations). Since these additional grains come from the "outside", all such systems can be described as *open*.

In a typical sand pile, the individual grains are considered to be perfect "cubes" which can be stacked on top of each other on an $L \times L$ grid. Various locations on the grid are uniquely specified by a pair of coordinates (x, y), where x and y are integers. In order to obtain reasonably complex behavior, it is necessary to introduce a "toppling rule", which determines when avalanches occur, and how the grains are to be redistributed. An obvious way to do this would be to assume that an avalanche begins when the height of the pile at some location (x, y) exceeds a given critical value Z_{cr}. At that point, one grain should be moved to each of the following four neighboring sites: $(x + 1, y)$, $(x - 1, y)$, $(x, y + 1)$ and $(x, y - 1)$.

To see how this simple rule can give rise to highly complex behavior, let us assume that we have a 5×5 grid, and that $Z_{cr} = 3$. As grains are added at random locations, the pile will initially grow without any "toppling" events, until the height reaches Z_{cr} at one or more points on the grid. What happens next will depend on the specific distribution of the grains. As an illustration, suppose that we have a situation such as the one shown in Fig. 10.1, where the height has reached the critical value at several locations. Such a system is clearly "poised" for an avalanche, since a redistribution of grains will almost certainly occur in the next few steps.

1	2	1	0	1
1	0	2	2	2
2	3	3	1	1
0	3	2	3	0
3	1	2	1	1

Figure 10.1: A sand pile that has reached the critical height in five locations.

If we assume that the next grain is added to the central square, we will observe an avalanche that involves as many as 6 toppling events at multiple sites (such an avalanche is said to have size $s = 6$). The various configurations that the system goes through before reaching a new stable state are shown in Fig. 10.2. In this figure, the color of the squares reflects the height of the pile at each location (the darker, the higher), and the circles indicate changes which have occurred in that particular step.

When analyzing the dynamic process described in Fig. 10.2, it is important to recognize that a different initial configuration could have resulted in a much smaller avalanche. If, for example, our starting distribution of grains happened to be the one shown in Fig. 10.3, the size of the avalanche would have been

10.1. SAND PILES AND AUTOCATALYSIS 207

$s = 2$ instead of $s = 6$ (the toppling events that occur in this case are indicated in Fig. 10.4).

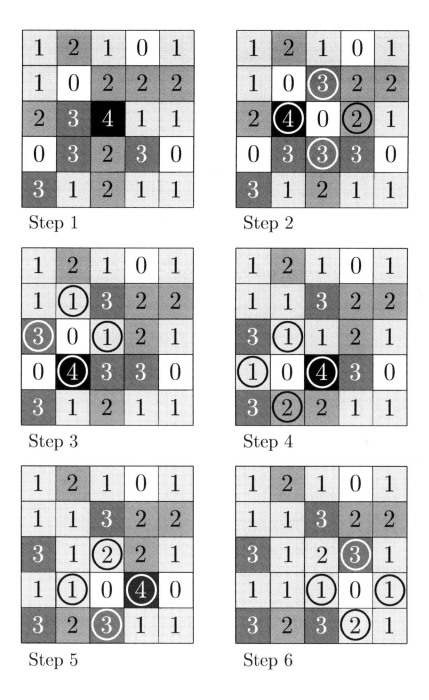

Figure 10.2: The sequence of toppling events for the configuration in Fig. 10.1.

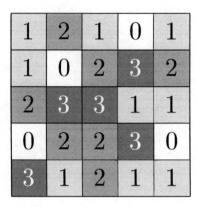

Figure 10.3: A sand pile with a different initial configuration.

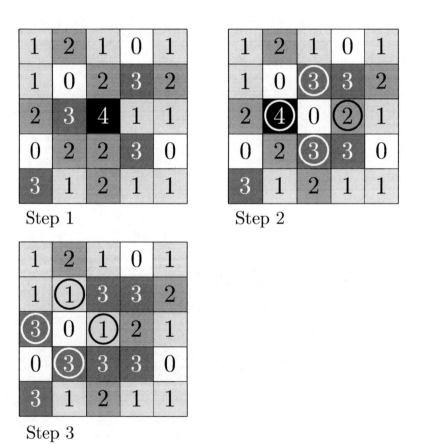

Figure 10.4: The sequence of toppling events for the configuration in Fig. 10.3.

10.1. SAND PILES AND AUTOCATALYSIS

One of the most important features of the sand pile model is its robustness with respect to parameter variations. Numerous experiments have shown that the underlying dynamics remain essentially the same if we modify the value of Z_{cr}, or change the location where new grains are placed.[20] In both cases, the system eventually returns to the critical state after every avalanche (such behavior is commonly referred to as "self-organized criticality"). This holds true even if we insert local "barriers" that prevent avalanches at certain points in the grid - these barriers may change the shape of the pile, but its dynamic behavior will remain "critical".

When we use the term "robust" to describe such models, we should emphasize that this is really a "meta-property" of the system. Indeed, although the overall behavior of a sand pile remains largely unaffected by parametric variations, the details of its evolution are not. These details are actually extraordinarily sensitive to the configuration of the system at the time when the next grain of sand is added. As a result, we cannot predict whether the next avalanche will be large, small or intermediate. Something similar holds true for many other complex systems, particularly those that exhibit some form of self-organization. In all such cases, it is virtually impossible to anticipate how the system will respond to a disturbance, even if it is very small.

Experiments performed on large sand piles have shown that the probability of observing an avalanche of size s satisfies a power law of the form

$$P(s) = s^{-\tau} \tag{10.1}$$

with exponent $\tau = 1.1$. It turns out that statistical descriptions of this sort (with different values of τ, of course) apply to many other types of complex systems as well, ranging from earthquakes to extinctions and floods. When interpreting what these laws mean, it is important to keep in mind that they have no predictive power. As a result, we have no way of anticipating the magnitude of the next event based on previous history – a pattern emerges only when we examine a large number of events that already occurred.

What is particularly interesting about power laws is the fact that they assign no special significance to "catastrophic" events. Although their consequences can be devastating, such major disruptions ultimately follow the same simple laws as smaller ones, and are "exceptional" only in the sense that they occur less frequently. When you think about it, this is a rather counterintuitive insight, since we normally assume that extraordinary occurrences require unusual causes which are not accounted for by standard models. Consider, for example, the extinction of dinosaurs, which many believe was caused by a large meteorite that struck the earth at some point in the distant past. What we have learned about complex systems in recent years suggests that this event may have occurred for reasons that are far less dramatic, and involve no major external factors. Those who promote such a view contend that the possibility of mass extinctions is actually "built into" the model, and requires no special circumstances.

Phenomena whose behavior conforms to power laws are also characterized by some form of *scale invariance,* in the sense that they have no features on any given scale that "stand out". We already encountered this property in the case of self-similar fractals (which also conform to power laws), but the principle is much broader, and can be viewed as one of the "trademarks" of complexity.

To put this claim into the proper perspective, we should recall that "typical" physical systems have a characteristic response time to perturbations, and a characteristic spatial length over which these effects are felt. Although random perturbations can lead to a range of different outcomes, the statistical distribution of key variables in such systems tends to be narrow, and the phenomenon can be adequately described by an "average" response.

It turns out, however, that this conclusion doesn't always apply to complex systems. When such a system reaches a certain critical threshold, a perturbation can lead to almost *any* kind of response. This means (among other things) that we have no reliable way of predicting how the system will react to a small disturbance. Under such circumstances, usual statistical measures such as the mean and variance are not very helpful, and are sometimes even impossible to define.

It is not difficult to see that this sort of behavior is consistent with power laws. As an illustration, suppose that we are observing a phenomenon which is characterized by a power law of the form

$$P(s) = Ks^{-\tau} \tag{10.2}$$

where $\tau > 1$. If s represents a measurable physical quantity, such a distribution is usually greater than zero only when s exceeds a certain threshold value (which we will denote in the following by s_1). In view of that, it would be reasonable to express $P(s)$ as

$$P(s) = \begin{cases} Ks^{-\tau} & (s \geq s_1) \\ 0 & (s < s_1) \end{cases} \tag{10.3}$$

The constant K that appears in (10.3) can be easily computed from the normalization requirement

$$\int_{-\infty}^{\infty} P(s)ds = \int_{s_1}^{\infty} Ks^{-\tau}ds = 1 \tag{10.4}$$

which can be rewritten as

$$K\int_{s_1}^{\infty} s^{-\tau}ds = \lim_{s\to\infty}\frac{K}{1-\tau}s^{1-\tau} - \frac{K}{1-\tau}\cdot s_1^{(1-\tau)} = -\frac{K}{1-\tau}\cdot s_1^{(1-\tau)} = 1 \tag{10.5}$$

10.1. SAND PILES AND AUTOCATALYSIS

Solving for K, we obtain

$$K = (\tau - 1) \cdot s_1^{(\tau-1)} \tag{10.6}$$

(note that K must be positive, since we assumed that $\tau > 1$).

Recalling that the mean value of s is given as

$$E[s] = K \int_{s_1}^{\infty} s \cdot (s^{-\tau}) ds = \lim_{s \to \infty} \frac{K}{2-\tau} s^{2-\tau} - \frac{K}{2-\tau} \cdot s_1^{2-\tau} \tag{10.7}$$

it follows that this concept is meaningless whenever $1 < \tau < 2$, because $E[s]$ becomes infinitely large for this range of values. If $2 < \tau < 3$, it is possible to define a mean value but not the standard deviation, since the variance

$$V(s) = E[s^2] - E[s] \tag{10.8}$$

has a diverging term

$$E[s^2] = K \int_{s_1}^{\infty} s^2 \cdot s^{-\tau} ds = \lim_{s \to \infty} \frac{K}{3-\tau} s^{3-\tau} \tag{10.9}$$

In such cases the mean value is of little use, since it is as likely to occur as *any* other value (in other words, it is not a "typical" feature of the system).

10.1.2 Autocatalysis

Another example of self-organization is the process of *autocatalysis*, which is essential for the proper operation of living cells. To get a sense for how this process works, consider a simple chemical reaction in which molecules A, B and C can combine to produce more complicated molecules AB, AC and BC. Schematically, this reaction can be described in the manner indicated in Fig. 10.5.

The circles in this graph represent the different molecules that can arise, and the squares symbolize the chemical reactions that produce them. The edges describe the nature of the reaction - incoming edges denote the "parent" molecules, and the outgoing edge points to the "product" of the reaction.

Given a set of molecules of type A, B and C, reactions that produce more complex molecules will occur infrequently, and even simple chains such as ABA or CAB will be in short supply. Things "speed up" considerably, however, if catalysts are present in the system. The catalysts could be external molecules D, E and F, whose role in the reaction is denoted by a dashed line in Fig. 10.6.

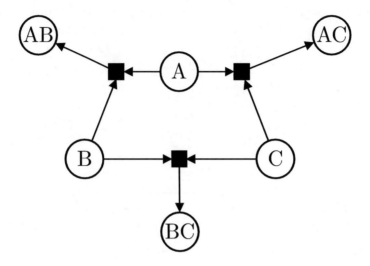

Figure 10.5: Reactions that create more complex molecules.

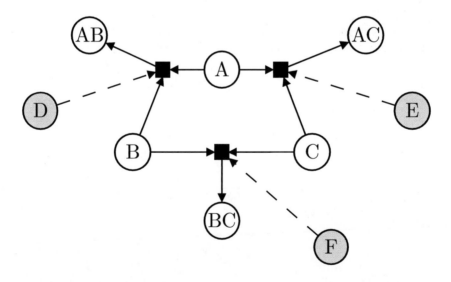

Figure 10.6: Reaction graph with external catalysts.

Alternatively, the process could be *autocatalytic*, in which case molecules that are already present in the system "accelerate" certain reactions. One such scenario is shown in Fig. 10.7, where the reactions that produce AB and AC are catalyzed from "within".

10.1. SAND PILES AND AUTOCATALYSIS

To understand the significance of autocatalytic processes, we first need to recognize that a chemical system cannot be self-sustaining unless the following two conditions are satisfied:

1) There must be a steady supply of initial molecules (like A and B in our example) from which all other molecules are formed. These initial molecules are sometimes referred to as "food molecules".

2) The system must be *open* - i.e., there must be a constant influx of matter and energy from the external environment.

The requirement that the system must be open is important because it allows for a broader range of possible configurations (including those that do not correspond to a "chemical equilibrium"). To see what this means, imagine what would happen if there was no influx of "food molecules" from the outside. Under such circumstances, it is likely that the system would ultimately settle into a state in which the two different types of molecules have similar concentrations. This would not be the case, however, if we were to introduce a surplus of A-type molecules into the system. Such a system would presumably have a steady-state as well, but the concentrations of A and B could be very different.

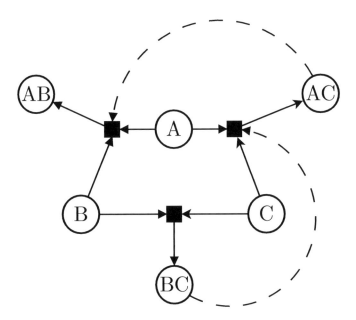

Figure 10.7: Example of an autocatalytic reaction.

A key property of reaction graphs such as the one in Fig. 10.7 is the ratio of the number of different molecule types (N) to the number of possible reactions (e). The importance of this ratio becomes apparent if we consider how one can build a graph starting from N isolated nodes. No matter how we choose to introduce new edges, the curve relating the size of the largest connected subgraph S_G to e/N will have the same general sigmoid form shown in Fig. 10.8. The sharp slope of this curve suggests that a large connected cluster will emerge as the number of edges approaches $N/2$. At that point, the system behavior changes abruptly and the likelihood of encountering complex molecules increases dramatically.

To get a better sense for how the ratio of edges to nodes affects reaction networks, let us consider a system in which there are three types of "food molecules", A, B and C. As more complex molecules begin to form, the number of possible reactions will gradually increase, as will the probability that some of them will be catalyzed. When this number reaches a certain critical threshold (relative to the number of different molecules), it is reasonable to expect that a large connected subgraph will emerge.

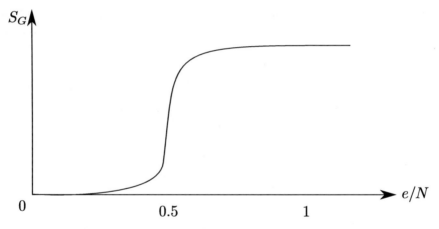

Figure 10.8: Size of the largest connected subgraph vs. e/N.

One such possibility is illustrated in Fig. 10.9, in which all catalyzed reactions are indicated by black squares (those that are not are colored gray). In this case, the autocatalytic set consists of molecules AC, BC, ACA and BCC, which are likely to be produced in larger quantities. It is important to note in this context that molecules ABB, ABAA and ACBCC are *not* part of this set, although they accelerate certain reactions within it.

When estimating the overall number of possible reactions, we should also take into account the fact that some of them might be reversible. This means that complex molecules such as ABAA could potentially disintegrate into AB and AA (or even simpler constituent parts). If we were to include this in our representation, each edge would become *bidirectional*, and their number would

10.2. SELF-ORGANIZATION AND INFORMATION

double. We could also consider interactions involving three or more molecules, which would increase the number of possible combinations even further. In both of these scenarios the number of reactions always grows much faster than the number of molecule types, so the emergence of an autocatalytic set becomes more likely.

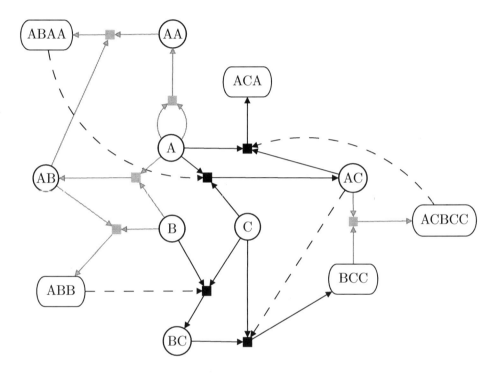

Figure 10.9: Example of an autocalystic set.

10.2 Self-Organization and Information

Complex behavior often requires close coordination between the individual "building blocks" that constitute the overall system. This is particularly evident in biological systems, which cannot function properly unless the underlying metabolic and physiological processes are perfectly synchronized. Since synchronization and coordination usually require some form of information exchange, it makes sense to briefly examine how complexity and self-organization might be described using the "language" of information theory.

We begin by observing that highly complex behavior rarely (if ever) occurs in closed systems, since they are subject to the Second Law of Thermodynamics. As a result, any form of initial organization that they may possess will eventually disappear as the system approaches thermodynamic equilibrium (which is a state of maximal "disorder"). Since such systems are incapable of preserving orderly patterns for long periods of time, their ability to generate and store information is minimal.

Open systems are quite different in that respect, since they allow for a constant exchange of matter and energy with their surroundings (which permits them to operate in states that are far removed from thermodynamic equilibrium). Such states can give rise to various forms of complex behavior, through which information can be created, processed and stored. In one way or another, all systems that are capable of self-organization exploit this possibility, regardless of whether they arise in nature or are engineered by humans.

A typical example of a man-made self-organizing physical system is a laser. The atoms in a laser are stimulated from the outside, which leads to the emission of photons. Initially, this process shows no signs of organized behavior, and the phases of the produced electromagnetic waves are completely uncorrelated. However, when the amplitude reaches a critical level, a coherent wave forms spontaneously. At that point, the corresponding electric field effectively "enslaves" the atoms, and forces them to oscillate in a particular way. When a laser is in this state, its dynamic behavior can be adequately described by a small number of macroscopic variables (just like a gas in thermodynamic equilibrium can be described in terms of pressure, volume and temperature). This implies an enormous compression of information, since it allows us to disregard the individual atomic states.

To illustrate the correlation between information and abrupt changes in the system behavior (which are considered to be a "trademark" of complexity), it is helpful to examine the operation of a laser in some more detail. In general, a laser can be viewed as an ensemble of N atoms, each of which can be in one of two possible energy states. The occupation probabilities of these two states can be expressed as

$$p_1 = \frac{N_1}{N} \tag{10.10}$$

and

$$p_2 = \frac{N_2}{N} \tag{10.11}$$

where N_1 and N_2 to denote the number of atoms that are in the *lower* and *higher* energy state, respectively. We can now use probabilities (10.10) and (10.11) to define the *information per atom* as

$$\bar{H} = -p_1 \ln p_1 - p_2 \ln p_2 \tag{10.12}$$

(which is consistent with the notion of information entropy that was introduced in Section 1.2). It is not difficult to see that the *total information entropy* for

10.2. SELF-ORGANIZATION AND INFORMATION

the entire ensemble can be expressed as

$$H = -N \left(p_1 \ln p_1 + p_2 \ln p_2 \right) \tag{10.13}$$

In order to calculate the probabilities p_1 and p_2 that appear in expression (10.13), we must relate them to some measurable physical quantities that characterize the operation of a laser. A suitable pair of quantities would be the number of photons at time t (denoted $n(t)$), and the so-called *inversion*, which is defined as [21]

$$D(t) = N_2(t) - N_1(t) \tag{10.14}$$

It turns out that the dynamic behavior of $n(t)$ and $D(t)$ can be described by a relatively simple second order model of the form

$$\begin{aligned} \dot{n} &= WDn - 2Kn \\ \dot{D} &= (D_0 - D)/T - 2WDn \end{aligned} \tag{10.15}$$

where W, K and T are constants. [22]

The constant T in equation (10.15) represents the relaxation time, which can be interpreted as the time required for $D(t)$ to reach its equilibrium value D_e. Since T tends to be a very small number in most practical models, the second equation in (10.15) can be approximated as

$$T\dot{D} = D_0 - D(t) - 2WTD(t)n(t) \approx 0 \tag{10.16}$$

This allows us to express $D(t)$ as

$$D(t) \approx \frac{D_0}{1 + 2WTn(t)} \approx D_0(1 - 2WTn(t)) \tag{10.17}$$

since $2WTn(t)$ is a small number. Substituting this into the equation for $\dot{n}$, we obtain

$$\dot{n} = WD(t)n(t) - 2Kn(t) = WD_0 \left[1 - 2WTn(t) \right] \cdot n(t) - 2Kn(t) =$$
$$= (WD_0 - 2K)n(t) - 2D_0W^2Tn^2(t) \tag{10.18}$$

Setting $\alpha = WD_0 - 2K$ and $\beta = 2D_0W^2T$, the overall model finally reduces to

$$\dot{n} = \alpha n(t) - \beta n^2(t) \tag{10.19}$$

It is important to recognize at this point that $n(t)$ must always be positive, since it represents the number of photons at time t. This means that we can have an increase in the production rate of photons (which corresponds to $\dot{n} > 0$)

only if $\alpha = WD_0 - 2K > 0$. Since this is a necessary condition for proper laser operation, we must ensure that the initial inversion D_0 satisfies

$$D_0 > \frac{2K}{W} \qquad (10.20)$$

In general, this can be accomplished by a sufficiently high level of "pumping".

In order to determine the equilibrium value for the number of photons, we need to set the left hand side of equation (10.18) to zero. This produces

$$0 = (WD_0 - 2K)n_e - 2D_0 W^2 T n_e^2 \qquad (10.21)$$

or equivalently

$$0 = n_e[(WD_0 - 2K) - 2D_0 W^2 T n_e] \qquad (10.22)$$

Equation (10.22) has two solutions - a trivial one ($n_e = 0$) and

$$n_e = \frac{WD_0 - 2K}{2D_0 W^2 T} \qquad (10.23)$$

Note that this second equilibrium exists only if

$$D_0 > \frac{2K}{W} \qquad (10.24)$$

since $n_e < 0$ makes no physical sense.

We can determine the equilibrium inversion, D_e, by using expression (10.17), which produces

$$D_e \approx D_0(1 - 2WT n_e) \qquad (10.25)$$

This equation allows for two possible scenarios.

CASE 1. If

$$D_0 \leq \frac{2K}{W} \qquad (10.26)$$

then $n_e = 0$ is the only feasible equilibrium, in which case $D_e \approx D_0$.

CASE 2. If inequality (10.20) holds, n_e takes the value given by (10.23). Substituting this expression into (10.25), we obtain

$$D_e \approx D_0 \left[1 - 2WT \cdot \frac{WD_0 - 2K}{2D_0 W^2 T}\right] = D_0 \left[1 - \frac{WD_0 - 2K}{D_0 W}\right] = \frac{2K}{W} \qquad (10.27)$$

Note that in this case the inversion ultimately settles into a value that is *lower* than D_0, since $WD_0 - 2K > 0$.

In order to determine how the information per atom changes as D_0 is increased, we should first observe that

$$-N \leq D_e \leq N \qquad (10.28)$$

10.2. SELF-ORGANIZATION AND INFORMATION

To see why this is so, it suffices to observe that in the equilibrium

$$D_e = N_2^e - N_1^e \tag{10.29}$$

and that

$$N = N_1^e + N_2^e \tag{10.30}$$

In the extreme case when $N_2^e = 0$, we have that $N = N_1^e$, and $D_e = -N_1^e = -N$. Similarly, if $N_1^e = 0$, we have $N = N_2^e$, and $D_e = N_2^e = N$.

From equations (10.29) and (10.30), we easily obtain N_1^e and N_2^e as

$$N_1^e = \frac{1}{2}(N - D_e) \tag{10.31}$$

and

$$N_2^e = \frac{1}{2}(N + D_e) \tag{10.32}$$

Dividing both expressions by N, we can now compute the equilibrium occupation probabilities for the two energy states as

$$p_1 = \frac{N_1^e}{N} = \frac{1}{2}(1 - \frac{D_e}{N}) \tag{10.33}$$

and

$$p_2 = \frac{N_2^e}{N} = \frac{1}{2}(1 + \frac{D_e}{N}) \tag{10.34}$$

Inserting these expressions into (10.12), the information per atom becomes

$$\bar{H} = -\frac{1}{2}(1 - \frac{D_e}{N})\ln\frac{1}{2}(1 - \frac{D_e}{N}) - \frac{1}{2}(1 + \frac{D_e}{N})\ln\frac{1}{2}(1 + \frac{D_e}{N}) \tag{10.35}$$

We can simplify expression (10.35) by setting

$$\gamma = \frac{D_e}{N} \tag{10.36}$$

which allows us to plot $\bar{H}$ as a function of γ (the corresponding graph is shown in Fig. 10.10).

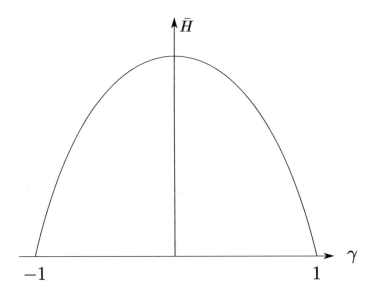

Figure 10.10: Information entropy $\bar{H}$ as a function of γ.

Some interesting additional insights can be gained by examining how $\bar{H}$ depends on D_0 (which is a quantity that we can control by adjusting the level of pumping). To see what kind of information can be extracted from this relationship, let us first recall that $n_e = 0$ whenever $D_0 \leq 2K/W$, which implies $D_e = D_0$. If D_0 exceeds $2K/W$, we have a new equilibrium $n_e \neq 0$ (given by equation (10.23)), which corresponds to a constant inversion $D_e = 2K/W$.

A schematic representation of this simple relationship is shown in Fig. 10.11, in which $D_c = 2K/W$ denotes the critical value when the system equilibrium changes. This graph suggests that the emergence of order corresponds to the "flat" part of the curve, where D_e (and therefore $\bar{H}$ as well) are independent of D_0.

10.2.1 The Maximum Information Principle

A more general (and deeper) connection between information and self-organization has to do with the so-called *Maximum Information Principle*, which is a powerful analytic tool for describing the dynamic behavior of complex systems.[23] In order to explain this principle, we first need to briefly describe conditions under which information entropy is maximized. As noted in Chapter 1, this quantity is defined as

$$H = -\sum_{j=1}^{M} p_j \ln p_j \qquad (10.37)$$

10.2. SELF-ORGANIZATION AND INFORMATION

where M is the number of possible messages, and p_j represents the probability that the source will generate message j.

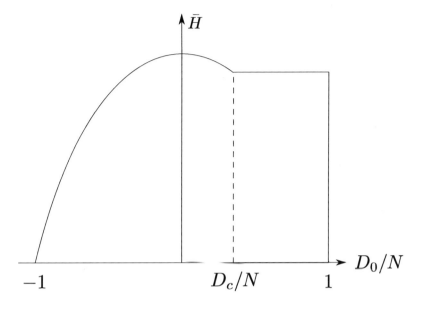

Figure 10.11: The onset of self-organized behavior.

With that in mind, our task will amount to maximizing H subject to constraint

$$\sum_{j=1}^{M} p_j = 1 \tag{10.38}$$

The following lemma shows how this can be done for a general function $f(x)$ given a set of algebraic constraints. What this result essentially tells us is that computing

$$\max_{x} f(x) \tag{10.39}$$

subject to

$$g(x) = 0 \tag{10.40}$$

is equivalent to the unconstrained maximization of function

$$\Phi(x, \lambda) = f(x) + \lambda^T g(x) = f(x) + \sum_{i} \lambda_i g_i(x) \tag{10.41}$$

where $\lambda = [\lambda_1 \, \lambda_2 \, \ldots \, \lambda_M]$ represent *Lagrange multipliers*.

Lemma 10.1. Let (x_0, λ_0) be the maximum of function $\Phi(x, \lambda)$ defined in (10.41). Then,

$$g(x_0) = 0 \tag{10.42}$$

and
$$f(x_0) > f(x) \tag{10.43}$$
over all possible x that satisfy equation (10.40).

Proof. Observe first that if (x_0, λ_0) is a maximum of $\Phi(x, \lambda)$, then
$$\left.\frac{\partial \Phi}{\partial \lambda_i}\right|_{x_0, \lambda_0} = g_i(x_0) = 0, \quad i = 1, 2, \ldots, M \tag{10.44}$$

Now, suppose that there exists some $x_1 \neq x_0$ such that
$$f(x_1) > f(x_0) \tag{10.45}$$
and
$$g(x_1) = 0 \tag{10.46}$$

In that case,
$$f(x_1) + \lambda_0^T g(x_1) > f(x_0) + \lambda_0^T g(x_0) \tag{10.47}$$
which implies
$$\Phi(x_1, \lambda_0) > \Phi(x_0, \lambda_0) \tag{10.48}$$
This is a contradiction, since (x_0, λ_0) is assumed to be a maximum of $\Phi(x, \lambda)$.
Q.E.D.

According to Lemma 10.1, maximizing
$$H = -\sum_{j=1}^{M} p_j \ln p_j \tag{10.49}$$
subject to
$$\sum_{j=1}^{M} p_j = 1 \tag{10.50}$$
can be reformulated as the *unconstrained* maximization of function
$$\Phi(p_1, \ldots, p_M, \lambda) = -\sum_{j=1}^{M} p_j \ln p_j + \lambda \left[\sum_{j=1}^{M} p_j - 1\right] \tag{10.51}$$
(note that in this case we have only one Lagrange multiplier, since (10.50) is the only constraint). It is easily verified that
$$\frac{\partial \Phi}{\partial p_k} = -\ln p_k - 1 + \lambda = 0 \quad (k = 1, 2, \ldots, M) \tag{10.52}$$

10.2. SELF-ORGANIZATION AND INFORMATION

from which we obtain
$$p_k = e^{\lambda-1} \quad (k = 1, 2, \ldots, M) \tag{10.53}$$

Substituting this into
$$\frac{\partial \Phi}{\partial \lambda} = \sum_{j=1}^{M} p_j - 1 = 0 \tag{10.54}$$

we get
$$\sum_{j=1}^{M} p_j = \sum_{j=1}^{M} e^{\lambda-1} = Me^{\lambda-1} = 1 \tag{10.55}$$

which implies that
$$e^{\lambda-1} = \frac{1}{M} \tag{10.56}$$

From equation (10.53), we can now conclude that information entropy is maximized when
$$p_1 = p_2 = \ldots = p_M = \frac{1}{M} \tag{10.57}$$

(i.e., when all probabilities are *equal*).

To see how the principle of maximizing information entropy can be applied to physical systems, let us consider the notion of *thermodynamic entropy*, which is measured by the number of different microscopic configurations that give rise to the *same* macroscopic state of a gas. Since the precise microscopic configuration cannot be identified from observable quantities such as pressure or temperature, there is always some uncertainty about the state of the gas on the molecular level. This uncertainty will obviously be *larger* if many different configurations correspond to the same macroscopic state, so it is fair to say that a state of maximal entropy is also the state in which our ignorance about the system is greatest.

In order to establish a more formal connection between thermodynamic entropy and information entropy, let us assume that the container is divided into M identical cells, each containing N_k molecules of the gas ($k = 1, 2, \ldots, M$). Since the molecules themselves are indistinguishable, the best way to characterize their spatial distribution is to provide the corresponding vector $[N_1\ N_2\ \ldots\ N_M]$. Note, however, that this vector *does not* tell us anything about which individual particles occupy a particular cell - it only gives us their aggregate number.

In statistical thermodynamics, the entropy of a macroscopic state can be calculated as
$$S = k_B \ln W(N_1, N_2, \ldots, N_M) \tag{10.58}$$

where k_B is Boltzmann's constant, and $W(N_1, N_2, \ldots, N_M)$ represents the number of different configurations that correspond to vector $[N_1\ N_2\ \ldots\ N_M]$. It is not difficult to see that we can place N_1 molecules in cell 1, N_2 molecules in cell 2, and so on in exactly

$$W(N_1, N_2, \ldots, N_M) = \frac{N!}{N_1! N_2! \ldots N_M!} \qquad (10.59)$$

different ways. For the sake of simplicity, we will demonstrate this for $M = 3$ (with no loss of generality).

We begin by observing that

$$W(N_1, N_2, N_3) = \binom{N}{N_1} \cdot \binom{N - N_1}{N_2} \cdot \binom{N - N_1 - N_2}{N_3} \qquad (10.60)$$

Recalling that $N - N_1 - N_2 = N_3$, (10.60) becomes

$$W(N_1, N_2, N_3) = \frac{N!}{N_1!(N - N_1)!} \cdot \frac{(N - N_1)!}{N_2!(N - N_1 - N_2)!} \cdot \binom{N_3}{N_3} \qquad (10.61)$$

which simplifies to

$$W(N_1, N_2, N_3) = \frac{N!}{N_1! N_2! N_3!} \qquad (10.62)$$

after the term $(N - N_1)!$ is canceled.

Since

$$S = k_B \ln W(N_1, N_2, \ldots, N_M) = k_B \ln \frac{N!}{N_1! N_2! \ldots N_M!} \qquad (10.63)$$

we can rewrite the entropy as

$$S = k_B [\ln N! - \ln N_1! - \ln N_2! - \ldots - \ln N_M!] \qquad (10.64)$$

Using Stirling's approximation (which is valid for $N > 100$), we can express $\ln N!$ as [24]

$$\ln N! \approx N(\ln N - 1) \qquad (10.65)$$

in which case (10.64) reduces to

$$S \approx k_B [N(\ln N - 1) - \sum_{i=1}^{M} N_i (\ln N_i - 1)] \qquad (10.66)$$

Let us now introduce the quantity $\bar{S} = S/N$, which can be interpreted as "normalized entropy". This quantity can be expressed as

$$\bar{S} = \frac{S}{N} \approx k_B [(\ln N - 1) - \sum_{i=1}^{M} \frac{N_i}{N} (\ln N_i - 1)] \qquad (10.67)$$

10.2. SELF-ORGANIZATION AND INFORMATION

which becomes

$$\bar{S} = k_B[\ln N - \sum_{i=1}^{M} \frac{N_i}{N} \ln N_i] \tag{10.68}$$

after appropriate cancellations are made.

In order to further simplify expression (10.68), we should note that

$$\ln N - \sum_{i=1}^{M} \frac{N_i}{N} \ln N_i = -\sum_{i=1}^{M} \frac{N_i}{N} \ln \frac{N_i}{N} \tag{10.69}$$

This identity follows directly from

$$-\sum_{i=1}^{M} \frac{N_i}{N} \ln \frac{N_i}{N} = -\sum_{i=1}^{M} \frac{N_i}{N} [\ln N_i - \ln N] =$$
$$= -\sum_{i=1}^{M} \frac{N_i}{N} \ln N_i + \sum_{i=1}^{M} \frac{N_i}{N} \ln N \tag{10.70}$$

after recognizing that the second term can be expressed as

$$\sum_{i=1}^{M} \frac{N_i}{N} \ln N = \frac{1}{N} \ln N \cdot \sum_{i=1}^{M} N_i = \ln N \tag{10.71}$$

Substituting (10.69) into (10.68), we obtain

$$\bar{S} = -k_B \sum_{i=1}^{M} \frac{N_i}{N} \ln \frac{N_i}{N} = -k_B \sum_{i=1}^{M} p_i \ln p_i \tag{10.72}$$

where

$$p_i = \frac{N_i}{N} \quad i = 1, 2, \ldots, M \tag{10.73}$$

Expression (10.72) indicates that the normalized thermodynamic entropy $\bar{S}$ has the same general form as information entropy, although probabilities p_i have a somewhat different meaning in this case (here they represent the likelihood that a molecule will be found in cell i). If we were now to maximize $\bar{S}$ subject to

$$\sum_{j} p_j = 1 \tag{10.74}$$

(following the procedure outlined in equations (10.51)-(10.57)), we would obtain

$$p_1 = p_2 = \ldots = p_M = \frac{1}{M} \tag{10.75}$$

which is consistent with the result that we derived for information entropy.

10.2.2 A General Framework

The simple example that we just examined suggests that we can define the information entropy of a physical system using expression (10.37) provided that we interpret $\{p_j\}$ as the probabilities that are associated with different microstates. In this section we will show how the Maximum Information Principle can help us determine these probabilities, given a set of macroscopic variables $\{X_1, X_2, \ldots, X_r\}$ whose mean values $\bar{x}_k$ ($k = 1, 2, \ldots, r$) are known. As we do so, we should bear in mind that this approach is fundamentally different from the one used in statistical thermodynamics, where we *start* from the probability distributions, and then use them to compute mean values of observable macroscopic quantities.

In the following, we will characterize microstate j by vector $[x_1^{(j)}\ x_2^{(j)}\ \ldots\ x_r^{(j)}]$, where $x_k^{(j)}$ represents the value of X_k that corresponds to this particular microstate. Using this notation, the expected value of X_k can be expressed as

$$\bar{x}_k = \sum_j p_j x_k^{(j)} \tag{10.76}$$

We now proceed to compute the probabilities $\{p_i\}$ by maximizing the information entropy

$$H = -\sum_i p_i \ln p_i \tag{10.77}$$

subject to

$$\sum_i p_i x_k^{(i)} - \bar{x}_k = 0 \quad (k = 1, 2, \ldots, M) \tag{10.78}$$

and

$$\sum_i p_i - 1 = 0 \tag{10.79}$$

Since this is a standard constrained optimization problem, we can solve it using the Lagrange multiplier method described in the previous section. As shown in equation (10.41), this amounts to finding a set of probabilities $p_1^*, \ldots, p_M^*$ and Lagrange multipliers $\lambda^*, \lambda_1^*, \ldots, \lambda_M^*$ for which the function

$$\begin{aligned}\Phi = -\sum_i p_i \ln p_i - \lambda_1 \left[\sum_i p_i x_1^{(i)} - \bar{x}_1\right] - \ldots \\ -\lambda_M \left[\sum_i p_i x_M^{(i)} - \bar{x}_M\right] - \lambda \left[\sum_i p_i - 1\right]\end{aligned} \tag{10.80}$$

10.2. SELF-ORGANIZATION AND INFORMATION

is maximal. We will now show that the probability distributions $\{p_1^*, \ldots, p_M^*\}$ can be expressed as

$$p_k^* = \exp\left[-\sum_j \lambda_j^* x_j^{(k)} - \lambda^* - 1\right] \quad (k = 1, 2, \ldots, M) \tag{10.81}$$

We begin by observing that the optimality condition

$$\frac{\partial \Phi}{\partial p_k} = 0 \quad (k = 1, 2, \ldots, M) \tag{10.82}$$

can be expressed as

$$\frac{\partial \Phi}{\partial p_k} = -\frac{\partial}{\partial p_k}\left[\sum_{i=1}^M p_i \ln p_i\right] - \lambda_1 x_1^{(k)} - \ldots - \lambda_M x_M^{(k)} - \lambda = 0 \tag{10.83}$$

Since

$$-\frac{\partial}{\partial p_k}\left[\sum_{i=1}^M p_i \ln p_i\right] = -\frac{\partial}{\partial p_k}(p_k \ln p_k) = -\ln p_k - 1 \tag{10.84}$$

we can rewrite (10.83) as a system of M equations of the form

$$-\ln p_k - \lambda_1 x_1^{(k)} - \ldots - \lambda_M x_M^{(k)} - \lambda - 1 = 0 \quad (k = 1, 2, \ldots, M) \tag{10.85}$$

Setting $\hat{\lambda} \equiv \lambda + 1$, it is easily verified that the probabilities can be expressed as

$$p_k = \exp\left[-\sum_j \lambda_j x_j^{(k)} - \hat{\lambda}\right] \quad (k = 1, 2, \ldots, M) \tag{10.86}$$

The Lagrange multipliers that appear in (10.85) can be determined from the second set of optimality conditions, which have the form

$$\frac{\partial \Phi}{\partial \lambda_k} = -\left[\sum_i p_i x_k^{(i)} - \bar{x}_k\right] = 0 \quad (k = 1, 2, \ldots, M) \tag{10.87}$$

and

$$\frac{\partial \Phi}{\partial \lambda} = -\left[\sum_i p_i - 1\right] = 0 \tag{10.88}$$

In order to solve these equations, we need to substitute the expressions for p_i given in (10.86), which produces

$$\sum_i \exp\left[-\sum_j \lambda_j x_j^{(i)} - \hat{\lambda}\right] \cdot x_k^{(i)} = \bar{x}_k \quad (k = 1, 2, \ldots, M) \tag{10.89}$$

and
$$\sum_i \exp\left[-\sum_j \lambda_j x_j^{(i)} - \hat{\lambda}\right] = 1 \qquad (10.90)$$

This is obviously a system of $M+1$ equations in $\hat{\lambda}, \lambda_1, \ldots, \lambda_M$, which can be solved explicitly given $\{\bar{x}_k\}$ and $\{x_k^{(i)}\}$. From these equations, we obtain $\{\hat{\lambda}^*, \lambda_1^*, \ldots, \lambda_M^*\}$, which can then be substituted into (10.86) to obtain the probability distributions $\{p_1^*, \ldots, p_M^*\}$ that correspond to the maximal information entropy, $H_{\max}$. The entropy itself can be calculated as

$$H_{\max} = -\sum_i p_i^* \ln p_i^* = -\sum_i p_i^* \left[-\sum_j \lambda_j^* x_j^{(i)} - \hat{\lambda}^*\right] \qquad (10.91)$$

or equivalently as

$$H_{\max} = \hat{\lambda}^* \sum_i p_i^* + \sum_i p_i^* \sum_j \lambda_j^* x_j^{(i)} \qquad (10.92)$$

Recalling that
$$\sum_i p_i^* = 1 \qquad (10.93)$$

and that
$$\sum_i p_i^* \sum_j \lambda_j^* x_j^{(i)} = \sum_j \lambda_j^* \sum_i p_i^* x_j^{(i)} = \sum_j \lambda_j^* \bar{x}_j \qquad (10.94)$$

we finally obtain
$$H_{\max} = \hat{\lambda}^* + \sum_{j=1}^{M} \lambda_j^* \bar{x}_j \qquad (10.95)$$

This expression is useful because it allows us to directly relate information entropy to the expected values of measurable macroscopic quantities.

Chapter 11

Notes and References

11.1 Notes for Chapters 8-10

1. See e.g. Hassan Khalil, *Nonlinear Systems*, Pearson, 2001.

2. It is important to recognize in this context that linearization provides only *local* information about an equilibrium. As a result, we can claim that trajectories originating close to $x^e = -2$ will be attracted by this equilibrium, but we can't say anything about other solutions of the system.

3. E. N. Lorenz, "Deterministic non-periodic flow", *Journal of Atmospheric Science*, **20**, pp. 130-141, 1963.

4. T. Matsumoto, "A chaotic attractor from Chua's circuit", *IEEE Transactions on Circuits and Systems*, **CAS-31**, pp. 1055-1058, 1984.

5. It is important to reiterate that expression (8.49) holds only if δx_0, δx_1, ... , δx_n are sufficiently small numbers. This effectively means that the separation between the solutions corresponding to x_0 and $x_0 + \delta x_0$ eventually approaches a finite upper bound, although it grows exponentially at the outset.

6. The irregularities in $x_1(t)$ are fairly subtle for this particular choice of p, but they become more pronounced if the parameter values is increased beyond 9.3.

7. It is interesting to note that the process described above is *irreversible*. Indeed, if we were to vary p from -4 to 4 and back, we would see that the

system equilibrium does *not* follow the same path. This is quite unlike the systems that we considered in Examples 8.3-8.6, whose behavior is reversible with respect to parametric changes.

8. Catastrophe theory claims that a cusp such as the one shown in Fig. 8.29 is the most complicated shape that the catastrophe set can take in a system with two parameters. This is not to say that there are no two-parameter models where this set is *not* a cusp. Such systems do exist, but they represent very special cases, and even the slightest perturbation in $f(x)$ will transform the catastrophe set into a collection of cusps.

9. It is interesting to note that the dynamic behavior shown in Fig. 9.8 is actually quite rare, since only 4 out of the 256 rules produce it. This holds true for random patterns as well (they can be generated in only 10 different ways).

10. Stephen Wolfram, *A New Kind of Science*, Wolfram Media, 2002.

11. The numbers in this diagram correspond to the decimal equivalent of the state which they represent. Thus, a configuration like BWW (or 100) is identified with 4, and WWWW (or 1111) with 15.

12. K. Mainzer and L. O. Chua, *The Universe as Automaton*, Springer, 2012.

13. It turns out that reversibility is a very rare property. Among the 256 possible cellular automata with two colors and nearest neighbor rules only six (or 2.3%) fall into this category. If we were to extend our analysis to the three color case (which allows for $7.6256 \cdot 10^{12}$ different rules), we would find that only 1,800 rules (or 0.000000236%) are reversible. A thorough examination of these two classes of automata shows that reversibility needn't necessarily be associated with simple patterns such as the one described in Table 9.4. There are actually a number of examples in the three color category that exhibit rather complex behavior.

14. L.O. Chua, S. Yoon and R. Dogaru, "A nonlinear dynamics perspective of Wolfram's New Kind of Science. Part I: Threshold of complexity", *International Journal of Bifurcation and Chaos*, **12**, pp. 2655–2766, 2002.

15. Ibid.

16. Note that the system shown in Fig. 9.21 actually has two more equilibria, but neither of these can be reached from $z_i(0) = 0$.

17. It is important to keep in mind that there are actually infinitely many choices of b_1, b_2, b_3, ρ_0, ρ_1, ρ_2 that can emulate Rule 110, since these parameters are assumed to be real numbers. The values used in our example should therefore be viewed as a sort of "representative" for an entire region in the six dimensional parameter space. This holds true for each of the 256 parameter choices that Chua used in his proof.

18. The idea was first introduced in: S. Kauffman, "Metabolic stability and epigenesis in randomly constructed genetic nets", *Journal of Theoretical Biology*, 22, pp. 437–467, 1969. More information on this subject can be found in: Stuart Kauffman, *The Origins of Order*, Oxford University Press, 1993.

19. This example was considered in: Stuart Kauffman, *At Home in the Universe*, Oxford University Press, 1996.

20. Per Bak, *How Nature Works*, Springer, 1996.

21. Note that $D(t) > 0$ means that the number of photons on the higher energy level exceeds the number of photons on the lower one, which is a state that can only be achieved by external excitation.

22. Hermann Haken, *Information and Self-Organization*, Springer, 2010.

23. Ibid.

24. See e.g.: Geoffrey Grimmett and David Stirzaker, *Probability and Random Processes*, Oxford University Press, 2001.

11.2 Further Reading

Chaos

1. John Nicolis, *Chaos and Information Processing*, World Scientific, 1991.

2. Steven Strogatz, *Nonlinear Dynamics and Chaos*, Perseus Books Publishing, 1994.

3. Robert Hilborn, *Chaos and Nonlinear Dynamics*, Oxford University Press, 2000.

4. A. Sengupta (Ed.), *Chaos, Nonlinearity, Complexity*, Springer, 2006.

5. Huaguang Zhang, Derong Liu and Zhiliang Wang, *Controlling Chaos*, Springer, 2009.

6. Aleksandar Zecevic, *The Unknowable and the Counterintuitive*, University Readers, 2012.

Catastrophe Theory

1. V. I. Arnold, *Catastrophe Theory*, Springer-Verlag, 1984.

2. Koncay Huseyin, *Multiple Parameter Stability Theory and its Applications: Bifurcations, Catastrophes, Instabilities*, Oxford University Press, 1986.

3. V. I. Arnold, V. S. Afraymovich Y. S. Il'yashenko and L. P. Shil'nikov, *Bifurcation Theory and Catastrophe Theory*, Springer, 1999.

4. Tim Poston and Ian Stewart, *Catastrophe Theory and Its Applications*, Dover, 2012.

5. Michel Demazure, *Bifurcations and Catastrophes: Geometry of Solutions to Nonlinear Problems*, Springer, 2013.

Cellular Automata

1. Andrew Ilachinski, *Cellular Automata*, World Scientific, 2001.

2. StephenWolfram, *A New Kind of Science*, Wolfram Media, 2002.

3. Joel Schiff, *Cellular Automata: A Discrete View of the World*, Wiley – Interscience, 2008.

11.2. FURTHER READING

4. A. Hoekstra, J. Kroc and P. Sloot (Eds.), *Simulating Complex Systems by Cellular Automata*, Springer, 2010.

5. K. Mainzer and L. O. Chua, *The Universe as Automaton*, Springer, 2012.

6. Teijiro Isokawa, Katsunobu Imai, Noboyuki Matsui, Ferdinand Peper and Hiroshi Umeo (Eds.), *Cellular Automata and Discrete Complex Systems*, Springer, 2015.

7. Andrew Adamatzky and Genaro Martinez (Eds.), *Designing Beauty: The Art of Cellular Automata (Emergence, Complexity and Computation)*, Springer, 2016.

Boolean Networks

1. Stuart Kauffman, *The Origins of Order*, Oxford University Press, 1993.

2. Stuart Kauffman, *At Home in the Universe*, Oxford University Press, 1996.

3. Ilya Shmulevich and Edward Dougherty, *Probabilistic Boolean Networks: The Modeling and Control of Gene Regulatory Networks*, SIAM, 2009.

4. Aleksandar Zecevic and Dragoslav Siljak, *Control of Complex Systems: Structural Constraints and Uncertainty*, Springer, 2010.

5. Daizhan Cheng, Hongsheng Qi and Zhiqiang Li, *Analysis and Control of Boolean Networks*, Springer, 2011.

Self-Organization

1. Robert Graham and Arne Wunderlin (Eds.), *Lasers and Synergetics: A Colloquium on Coherence and Self-Organization in Nature*, Springer, 1987.

2. Per Bak, *How Nature Works*, Springer, 1996.

3. Henrik Jensen, *Self-Organized Criticality*, Cambridge University Press, 1998.

4. Scott Camazine, Jean-Louis Deneubourg, Nigel Franks, James Sneyd, Guy Theraula and Eric Bonabeau, *Self-Organization in Biological Systems*, Princeton University Press, 2003.

5. Didier Sornette, *Critical Phenomena in Natural Sciences: Chaos, Fractals, Self-Organization and Disorder*, Springer, 2009.

6. Cyrille Bertelle, Gerard Duchamp, Hakima Kadri-Dahmani (Eds.), *Complex Systems and Self-Organization Modelling*, Springer, 2009.

7. Hermann Haken, *Information and Self-Organization*, Springer, 2010.

8. Markus Aschwanden, *Self-Organized Criticality in Astrophysics: The Statistics of Nonlinear Processes in the Universe*, Springer, 2011.

Part IV
Mathematical Infinity

Chapter 12

Numbers, Limits and Infinite Sums

We all have experiences of processes that are "open-ended". Counting is a perfect example – we register it as a series of discrete events, but we tend to think of it as something that can continue indefinitely (although we are not able to do so ourselves). The capacity to think of such processes as "endless" is what has allowed us to develop the concept of infinity, and apply it both in science and in mathematics.

In order to properly internalize this idea, we normally rely on a cognitive metaphor that is sometimes referred to as the *Basic Metaphor of Infinity* (or BMI for short).[1] This metaphor provides us with a way to "complete" open-ended iterative processes by assuming that they have an *actual result*, and that this result is *unique*.

Since these two assumptions are obviously extensions of properties that we routinely encounter when dealing with finite iterative processes, they represent abstract creations of the human mind that have no concrete basis in physical reality. Nevertheless, there is no doubt that they provide us with a very powerful conceptual tool, which has allowed us to develop theoretical models that have greatly advanced our understanding of natural phenomena. In this chapter (and in the next one), we will take a closer look at some of these models, and examine the different ways in which the notion of infinity appears in mathematics.

12.1 Integers, Primes and Real Numbers

12.1.1 Prime Numbers

It makes sense to begin our discussion with number theory, since the concept of infinity has its roots in counting. Prime numbers play a particularly important role in this theory, since every integer can be uniquely represented as a product of its prime factors. By virtue of this property, we can think of primes as the fundamental "building blocks" from which all integers are formed. The following theorem demonstrates that there are infinitely many such numbers (it was proved by Euclid in the 4th century B.C.)

Theorem 12.1. There are infinitely many prime numbers.

Proof. Let us consider an arbitrary finite list of prime numbers, and assume that the largest number in the list is p. We now proceed to show that there is always a prime number that is greater than p, regardless of how large p happens to be. In order to do that, let us form an integer

$$N = (2 \cdot 3 \cdot 5 \cdot \ldots \cdot p) + 1 \tag{12.1}$$

and consider the following two (mutually exclusive) alternatives.

1. If we assume that N is a prime number, (12.1) implies that it is clearly larger than p.

2. If we assume that N is *not* a prime number, we can always express it as a product of its prime factors. However, from (12.1) we see that N cannot be divided by 2, 3, 5, ... , p without a remainder. As a result, if it really does have prime factors, then each of these numbers must be larger than p.

Both scenarios obviously imply that there exists a prime number greater than p. Since p was chosen *arbitrarily*, it directly follows that the set of prime numbers has infinitely many elements. **Q.E.D.**

One of the most interesting open problems related to prime numbers has to do with so-called "twin primes". Twin primes are numbers that appear in clusters of two, and have the general form $(p, p+2)$ (obvious examples are $(3, 5)$, $(11, 13)$, $(17, 19)$ and so on). What remains unclear is whether or not the number of such pairs is finite. Mathematicians have found some very large twin primes (such as $3,756,801,695,685 \cdot 2^{666,669} - 1$ and $3,756,801,695,685 \cdot 2^{666,669} + 1$, for example), but we do not know whether even larger ones exist.

Another question that remains unanswered is whether prime numbers can be generated in a systematic manner (and if so, how). Part of the problem lies in the fact that their distribution along the number line is *not monotonic*. There are, for example, 25 of them between 1 and 100, 21 between 100 and 200, 16 between 200 and 300, but there are 17 primes between 400 and 500. It turns out, however, that the *density* of primes on interval $[0 \ n]$ exhibits a hidden statistical regularity as $n \to \infty$. In particular, it can be shown that

$$\pi(n) \approx \frac{n}{\ln n} \tag{12.2}$$

when n is sufficiently large (where $\pi(n)$ denotes the number of primes that are less than or equal to n). This result is known as the Prime Number Theorem.

12.1.2 Real Numbers

When you think about it, there is actually nothing particularly "real" about real numbers. Since these numbers contain infinitely many decimals, they are clearly not something that we can describe precisely. The best we can do is to conceptualize such entities by resorting to metaphors which link them to finite objects and processes that we are familiar with.

There are actually three equivalent ways in which this can be done, each of which involves a somewhat different extrapolation from the finite to the infinite. The first one asks us to imagine a sequence of sets that are constructed in the following manner:

$$
\begin{aligned}
R_1 &= \{x.0,\ x.1,\ x.2, \ldots,\ x.9\} \\
R_2 &= \{x.00,\ x.01, \ldots, x.10, \ldots,\ x.90, \ldots, x.99\} \\
R_3 &= \{x.000,\ x.001, \ldots, x.100, \ldots,\ x.900, \ldots, x.999\}
\end{aligned}
\tag{12.3}
$$

$$\vdots$$

where x denotes an arbitrary integer. The set R_n that is obtained in the n-th step of this process will consist of all rational numbers in the interval $[x\ \ x+1]$ whose decimal expansion contains exactly n digits. The BMI metaphor that we introduced at the beginning of this chapter allows us to complete this sequence, and conceptualize set R_∞ in which each element has *infinitely* many decimals (and is therefore *not* a rational number).[2]

The second way to introduce real numbers is purely formal, and is based on 10 fundamental axioms that define their properties. These axioms make no reference to our intuition, and are therefore completely autonomous. We should note, however, that only one of them draws a clear distinction between real and rational numbers. This is the so-called *Least Upper Bound Axiom*, which stipulates that every nonempty set that has an upper bound also has a least upper bound.

What exactly is a "least upper bound"? To see that, let us consider a set of rational numbers $S_n = \{x_1, x_2, \ldots, x_n\}$ in which x_k consists of k decimals, and is formed by adding a single digit to number x_{k-1} at the k-th decimal place. An example of such a set with $n = 6$ would be $S_6 = \{1.2, 1.23, 1.236, 1.2368, 1.23680, 1.236802\}$. It is not difficult to see that the elements of such a set satisfy satisfy the condition $x_1 \leq x_2 \leq \ldots \leq x_n$ by construction, and that number x_n constitutes its least upper bound (in the sense that it is the *smallest* number that is greater than or equal to every member of S_n).

What happens to such a set when we let $n \to \infty$? Does a least upper bound still exist, and if so, what are its properties? The BMI metaphor allows us to assume that it does exist, and that it is unique. Such a number cannot be rational, however, since it has infinitely many decimals. As a result, if the Least Upper Bound Axiom is to remain in effect for infinite sets, we must introduce real numbers into the system.

A third way to think of real numbers involves the construction of successively smaller intervals. To see how this works, we will need to consider pairs of rational numbers (x, y) that differ by one in their last decimal digit. A sequence of such intervals $[x_n, y_n]$ is shown in Table 12.1.

$$[2.1, 2.2]$$
$$[2.13, 2.14]$$
$$[2.135, 2.136]$$
$$[2.1358, 2.1359]$$
$$\vdots$$

Table 12.1. A sequence of successively smaller intervals.

It is not difficult to see that the interval lengths become smaller in each step, and that $\{x_n\}$ is an increasing sequence, while sequence $\{y_n\}$ decreases. The BMI allows us to assume the length of these intervals reduces to zero when $n \to \infty$, and that the final result of this process is a unique number $r = x_\infty = y_\infty$. Once again, this number cannot be rational, since it has infinitely many digits in its decimal expansion by construction.

How are Real Numbers Approximated?

There is an entire field within number theory that is devoted to the problem of approximating real numbers by rational numbers (such approximations are known Diophantine approximations, in honor of the 3rd century mathematician Diophantus of Alexandria). Some of the key results in this area of mathematics are based on the recognition that all real numbers fall into one of two general categories - a number is said to be *algebraic* if it is a root of a polynomial with integer coefficients, and it is *transcendent* if this is not the case.

From a practical standpoint, approximating algebraic numbers is fairly simple, and can be done using Newton's method (which is capable of producing as many decimals as we choose). The following two examples illustrate how this process works.

Example 12.1. The number $\phi = (1 + \sqrt{5})/2 = 1.61803...$ (which is commonly referred to as the Golden Ratio) has had a special significance in Western culture. An efficient way to approximate it would be to iteratively solve the quadratic equation

$$F(x) = x^2 - x - 1 = 0 \tag{12.4}$$

12.1. INTEGERS, PRIMES AND REAL NUMBERS

since ϕ represents one of its roots. We can do so using Newton's method

$$x_{n+1} = x_n - [F'(x_n)]^{-1} F(x_n) \tag{12.5}$$

which produces the sequence

$$x_{n+1} = x_n - \frac{(x_n^2 - x_n - 1)}{2x_n - 1} \tag{12.6}$$

Given an appropriate initial guess, this sequence rapidly approaches ϕ, as illustrated in Table 12.2 (if we pick $x_0 = 1$, we obtain the first 8 digits in the decimal expansion after only 4 iterations). However, if we were to choose $x_0 = 0$, the sequence would converge to $x^* = -0.61803$, which is the *other* root of equation (12.4). We should therefore exercise some care in using this approach, since it needn't always produce the number that we are looking for.[3]

$$x_0 = 1$$
$$x_1 = 2$$
$$x_2 = 5/3 = 1.6\bar{6}$$
$$x_3 = 34/21 = 1.619047619$$
$$x_4 = 1,597/987 = 1.618034448$$
$$\vdots$$

Table 12.2. Successive approximations of ϕ using Newton's method.

Example 12.2. Another example of how algebraic numbers can be approximated is $\sqrt{2}$, which is irrational but satisfies the quadratic equation

$$F(x) = x^2 - 2 = 0 \tag{12.7}$$

Observing that $F'(x) = 2x$ and applying Newton's method, the iterative sequence

$$x_{n+1} = x_n - [F'(x_n)]^{-1} F(x_n) \tag{12.8}$$

becomes

$$x_{n+1} = x_n - \frac{(x_n^2 - 2)}{2x_n} = \frac{x_n}{2} + \frac{1}{x_n} \tag{12.9}$$

Given an appropriate initial guess, this sequence rapidly approaches $\sqrt{2}$, as illustrated in Table 12.3.

Transcendental numbers are generally much more difficult to approximate, since most of them cannot be described in terms of simple iterative procedures.

There are, however, some notable exceptions, the most "famous" being e and π. Number $e = 2.71828...$ (which is the base of natural logarithms) is defined as
$$e = \lim_{n \to \infty} (1 + \frac{1}{n})^n \qquad (12.10)$$
and can therefore be approximated by the sequence
$$x_n = (1 + \frac{1}{n})^n \qquad (12.11)$$
to an arbitrary precision. The first five terms of this sequence are shown in Table 12.4 (note that all of them are rational numbers whose numerators and denominators increase monotonically).

$x_0 = 1$

$x_1 = 3/2 = 1.5$

$x_2 = 17/12 = 1.416\bar{6}$

$x_3 = 577/408 = 1.414215686$

$x_4 = 665,857/470,832 = 1.414213562$

$\vdots$

Table 12.3. Successive approximations of $\sqrt{2}$.

$x_0 = 1$

$x_1 = 2$

$x_2 = 9/4 = 2.25$

$x_3 = 64/27 = 2.370370$

$x_4 = 625/256 = 2.441406$

$x_5 = 7,776/3,125 = 2.48832$

$\vdots$

Table 12.4. Successive approximations of e.

An alternative way to approximate e is based on the Taylor series expansion of function e^x, which is given as
$$e^x = \sum_{n=0}^{\infty} \frac{x^n}{n!} \qquad (12.12)$$

12.1. INTEGERS, PRIMES AND REAL NUMBERS

Setting $x = 1$ in (12.12) produces

$$e = \sum_{n=0}^{\infty} \frac{1}{n!} \qquad (12.13)$$

which is once again a sequence of rational numbers. The convergence rate of this sequence can be estimated from Table 12.5, which indicates that expression (12.13) is a more efficient way to approximate e than (12.10).

$$x_0 = 1$$
$$x_1 = 2$$
$$x_2 = 5/2 = 2.5$$
$$x_3 = 16/6 = 2.6\bar{6}$$
$$x_4 = 65/24 = 2.708\bar{3}$$
$$x_5 = 326/120 = 2.71\bar{6}$$
$$\vdots$$

Table 12.5. Approximations of e using the Taylor series expansion.

The formulas for approximating π are more diverse, since this number has a much longer history than e. One of the oldest ones is the infinite sum

$$\frac{\pi}{4} = \sum_{n=0}^{\infty} \frac{(-1)^n}{2n+1} \qquad (12.14)$$

which was proposed by Leibniz in the 17th century (based on an earlier result derived by James Gregory). This expression can be easily obtained by setting $x = 1$ into the Taylor series expansion

$$\tan^{-1}(x) = x - \frac{x^3}{3} + \frac{x^5}{5} - \frac{x^7}{7} + \cdots \qquad (12.15)$$

The difficulty with the Leibniz-Gregory formula is that it converges very slowly - it takes more than 600 terms, for example, to accurately reproduce the first two decimals of π. A much more efficient approximation is associated with the so-called *Basel Problem*, which was solved by Swiss mathematician Leonhard Euler in 1734. This problem involves the sum

$$S = \sum_{n=1}^{\infty} \frac{1}{n^2} = 1 + \frac{1}{4} + \frac{1}{9} + \frac{1}{16} + \cdots \qquad (12.16)$$

whose value was unknown prior to Euler's discovery (he showed that $S = \pi^2/6$).

An elegant way to prove of Euler's result is outlined below (this is not the approach that Euler used, but it is rigorous and quite straightforward). We begin by observing that a periodic function $f(x)$ whose period is $T = 2\pi$ can be expressed by a Fourier series of the form

$$f(x) = \frac{a_0}{2} + \sum_{n=1}^{\infty} a_n \cos(nx) + \sum_{n=1}^{\infty} b_n \sin(nx) \qquad (12.17)$$

where

$$a_0 = \frac{1}{\pi} \int_{-\pi}^{\pi} f(x)dx \qquad (12.18)$$

$$a_n = \frac{1}{\pi} \int_{-\pi}^{\pi} f(x) \cos(nx) dx \qquad (12.19)$$

and

$$b_n = \frac{1}{\pi} \int_{-\pi}^{\pi} f(x) \sin(nx) dx \qquad (12.20)$$

In its complex variant, this series takes the form

$$f(x) = \sum_{n=-\infty}^{\infty} c_n e^{inx} \qquad (12.21)$$

where

$$c_n = \frac{1}{2\pi} \int_{-\pi}^{\pi} f(x) e^{-inx} dx \qquad (12.22)$$

Let us now consider the special case when $f(x)$ has the form shown in Fig. 12.1. It is not difficult to see that this function has period 2π, and that $f(x) = x$ on interval $[-\pi \ \pi]$.

12.1. INTEGERS, PRIMES AND REAL NUMBERS

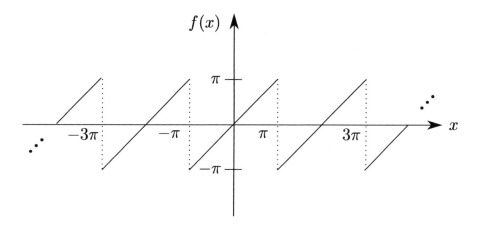

Figure 12.1: A periodic extension of function $f(x) = x$.

The coefficients c_n that correspond to $f(x)$ can be computed as

$$c_n = \frac{1}{2\pi} \int_{-\pi}^{\pi} x e^{-inx} dx = \frac{(-1)^n}{n} i \qquad (12.23)$$

and according to Parseval's Theorem, these coefficients must satisfy

$$\sum_{n=-\infty}^{\infty} |c_n|^2 = \frac{1}{2\pi} \int_{-\pi}^{\pi} |f(x)|^2 dx \qquad (12.24)$$

Observing that

$$|c_n|^2 = \frac{1}{n^2} \qquad (12.25)$$

for $n \neq 0$ and recognizing that

$$c_0 = \frac{1}{2\pi} \int_{-\pi}^{\pi} x\, dx = 0 \qquad (12.26)$$

we obtain

$$\sum_{n=-\infty}^{\infty} |c_n|^2 = 2 \sum_{n=1}^{\infty} \frac{1}{n^2} \qquad (12.27)$$

(in (12.27) we used the symmetry of n^2 with respect to sign changes). Substituting this into equation (12.24), we obtain

$$2 \sum_{n=1}^{\infty} \frac{1}{n^2} = \frac{1}{2\pi} \int_{-\pi}^{\pi} x^2 dx = \frac{\pi^2}{3} \qquad (12.28)$$

which implies that[4]

$$S = \sum_{n=1}^{\infty} \frac{1}{n^2} = \frac{\pi^2}{6} \qquad (12.29)$$

12.2 Division by Zero

Division by zero seems to be yet another way in which infinity makes its appearance in mathematics. Indeed, in the literature one still occasionally encounters expressions such as $1/0 = \infty$ or $1/\infty = 0$, which suggest that infinity lends itself to standard algebraic manipulations. It is important to keep in mind, however, that such operations have no meaning, since 0 and 1 are *numbers*, while infinity is best understood as a *concept*. To illustrate some of the difficulties that arise when this distinction is not observed, consider the expression

$$x = \frac{1}{0} \qquad (12.30)$$

If this were a legitimate algebraic operation involving numbers, it would follow that

$$0 \cdot x = 1 \qquad (12.31)$$

However, we know that

$$0 \cdot x = 0 \qquad (12.32)$$

by definition, so we arrive at the absurd conclusion that $0 = 1$. For similar reasons, it makes no sense to consider expressions such as

$$x = \frac{0}{0} \qquad (12.33)$$

since

$$0 \cdot x = 0 \qquad (12.34)$$

holds for *any* choice of x. In this case, we are forced to rule out division by zero because it fails to produce a unique answer.

Difficulties of this sort can be avoided by using *limits*, which allow us to work with functions such as

$$f(x) = \frac{\sin x}{x} \qquad (12.35)$$

and define their value at $x = 0$ as

$$f(0) = \lim_{x \to 0} f(x) \qquad (12.36)$$

By doing so, we avoid dealing with meaningless algebraic expressions of the form

$$f(0) = \frac{\sin 0}{0} = \frac{0}{0} \qquad (12.37)$$

12.2. DIVISION BY ZERO

since the limit is *never actually reached* (although we can come arbitrarily close to it).

The easiest way to analyze how the function defined in (12.35) behaves in the neighborhood of zero is to make use of the Taylor series expansion of $\sin x$, which has the form

$$\sin x = x - \frac{1}{3!} x^3 + \frac{1}{5!} x^5 - \frac{1}{7!} x^7 + \ldots \tag{12.38}$$

This expansion (which involves infinitely many terms, and is therefore a limit in its own right) allows us to express $f(x)$ as

$$f(x) = \frac{\sin x}{x} = 1 - \frac{1}{3!} x^2 + \frac{1}{5!} x^4 - \frac{1}{7!} x^6 + \ldots \tag{12.39}$$

which yields

$$f(0) = \lim_{x \to 0} \frac{\sin x}{x} = 1 \tag{12.40}$$

We encounter a similar situation with functions such as

$$f(x) = \frac{1}{x^2} - \frac{\cos x}{x^2} \tag{12.41}$$

which take the form

$$f(0) = \infty - \infty \tag{12.42}$$

if $x = 0$ is substituted directly. If ∞ were a number, one would expect that expression (12.42) should produce zero, since we are subtracting a number from itself. It is easily verified, however, that this is not the case. To see why, it suffices to observe that the Taylor series expansion of $\cos x$ has the form

$$\cos x = 1 - \frac{1}{2!} x^2 + \frac{1}{4!} x^4 - \ldots \tag{12.43}$$

As a result, we have that

$$1 - \cos x = \frac{1}{2!} x^2 - \frac{1}{4!} x^4 + \ldots \tag{12.44}$$

which implies that $f(x)$ can be expressed as

$$f(x) = \frac{1 - \cos x}{x^2} = \frac{1}{2!} - \frac{1}{4!} x^2 + \ldots \tag{12.45}$$

Such a representation of $f(x)$ allows us to evaluate this function for all values of x, including $x = 0$. Since $f(0) = 1/2$, we can conclude that in this particular case

$$f(0) = \infty - \infty \neq 0 \tag{12.46}$$

which clearly indicates that infinity is *not* a number.

The problem of dividing by zero also arises in the calculation of derivatives. Newton resolved this difficulty by defining the derivative of function $f(x)$ as

$$f'(x) = \frac{df}{dx} = \lim_{\Delta x \to 0} \frac{f(x + \Delta x) - f(x)}{\Delta x} \tag{12.47}$$

which allowed him to avoid dividing zero by itself (since the limit is never actually reached). It is interesting to note in this context that Leibniz used a very different approach, which involves quantities known as *infinitesimals*. At that time (and for several centuries to come), infinitesimals were considered to be highly problematic, because they were assumed to be smaller than any number, and yet greater than zero. This apparent paradox led the 18th century bishop (and philosopher) George Berkeley to remark that: "[Infinitesimals] are neither finite quantities, nor quantities infinitely small, nor yet nothing. May we not call them the ghosts of departed quantities?"[5]

The notion of an "infinitesimal" represents an extension of our everyday experience, which suggests that adding something very small to something very large produces no perceptible difference. Leibniz was the first mathematician to systematically develop this idea, and propose that an infinitely small interval might have an "objective existence". Unlike Newton, Leibniz interpreted df/dx as the ratio of two *actual numbers*, df and dx. Since neither of these two numbers was zero in his version of calculus, he was able to divide one by the other without encountering any conceptual problems. Newton's approach eventually prevailed, but the ideas that Leibniz developed resurfaced in the 1960s and 70s, when a series of new advances in set theory showed that infinitesimals have a legitimate place in the world of mathematics.

To get a sense for how Leibniz resolved the derivative problem, let ∂ denote a number that is greater than zero, but is smaller than any real number (which is how infinitesimals are defined). We will assume that algebraic operations apply to these numbers, which allows us to define quantities such as 5∂ and $2\partial^3$ (and therefore form polynomials in ∂ as well). In dealing with numbers of this sort, it is useful to introduce an operator that can "filter them out" when necessary, and ensure that the obtained results are real numbers. Such an operator is known as the *real approximation operator*, $\hat{R}$, which can transform expressions such as $5 + 2\partial - \partial^2$ into $\hat{R}(5 + 2\partial - \partial^2) = 5$.

Operator $\hat{R}$ allows us to define derivatives without using limits, as

$$f'(x) = \hat{R}\left(\frac{f(x + \partial) - f(x)}{\partial}\right) \tag{12.48}$$

As an illustration of how this works in practice, let us consider the derivative

12.3. INFINITE SUMS

of function $f(x) = x^2$. According to (12.48), $f'(x)$ can be computed as

$$f'(x) = \hat{R}\left(\frac{(x+\partial)^2 - x^2}{\partial}\right) = \hat{R}\left(\frac{2x\partial + \partial^2}{\partial}\right) = \hat{R}(2x + \partial) = 2x \quad (12.49)$$

which is exactly the same result obtained using the standard procedure. The advantage in using (12.48) is that it requires no limits, and utilizes only basic arithmetic.

The example that we just considered suggests that working with infinitesimals provides us with a "finer" picture of what actually happens in the process of differentiation, since the intermediate result $f'(x) = 2x + \partial$ contains more information about the derivative of $f(x) = x^2$ than $f'(x) = 2x$ (which is just an approximation). On the other hand, much of the elegance of traditional calculus would be lost if we insisted on including infinitesimals in all our computations, since that would make them considerably more complicated. The operator $\hat{R}$ allows us to avoid this difficulty, and disregard such numbers whenever they begin to compromise the efficiency of the calculation.

12.3 Infinite Sums

Infinite sums have attracted mathematicians for centuries, and continue to be a topic of significant interest. It is clear, of course, that such sums can never be computed in practice, which is why we define

$$S = \sum_{k=1}^{\infty} a_k \quad (12.50)$$

as the limit

$$S = \lim_{n \to \infty} S_n \quad (12.51)$$

where $\{S_n\}$ represent a sequence of partial sums of the form

$$S_n = \sum_{k=1}^{n} a_k \quad (12.52)$$

Since each of these partial sums can be evaluated precisely, we can examine whether or not sequence $\{S_n\}$ converges and what its limit might be.

Such an approach seems pretty straightforward, but implementing it in practice is not. Even some of the simplest cases can be rather misleading, since a sequence of partial sums that grows very slowly may end up diverging as $n \to \infty$. The following three examples illustrate just how "tricky" it can be to determine whether or not an infinite sum has a limit.

Example 12.3. One of simplest and most "innocent looking" infinite sums is the so-called *harmonic series*, which has the form

$$S = \sum_{k=1}^{\infty} \frac{1}{k} \qquad (12.53)$$

Empirical considerations suggest that S ought to be a finite number, since summing up the first 10^{43} terms produces $S_n \approx 100$ (which is a small number considering how many terms have been included). It turns out, however, that sequence $\{S_n\}$ actually *diverges*, although it would take an extraordinarily large amount of computation to establish this by simulation.

The following proof (which is due to the 14th century French mathematician Nicolas Oresme) provides some insight into why this is the case. What is remarkable about it is that it was derived nearly 400 years before the concept of a limit was properly defined.

We begin by observing that the terms in (12.53) can be rewritten as

$$S = 1 + \frac{1}{2} + \sum_{k=1}^{\infty} X_k \qquad (12.54)$$

where

$$\begin{aligned} X_1 &= 3^{-1} + 4^{-1} \\ X_2 &= 5^{-1} + 6^{-1} + 7^{-1} + 8^{-1} \\ X_3 &= 9^{-1} + 10^{-1} + \ldots + 16^{-1} \\ &\vdots \\ X_k &= (2^k + 1)^{-1} + (2^k + 2)^{-1} + \ldots + (2^k + 2^k)^{-1} \\ &\vdots \end{aligned} \qquad (12.55)$$

By construction, X_k contains 2^k elements, the *smallest* of which is $(2^k + 2^k)^{-1} = 2^{-(k+1)}$. This implies that

$$X_k > 2^k \cdot 2^{-(k+1)} = \frac{1}{2} \qquad (12.56)$$

for all k. Summing up n such terms, we obtain

$$S_n = \sum_{k=1}^{n} X_k > \sum_{k=1}^{n} \left(\frac{1}{2}\right) = \frac{n}{2} \qquad (12.57)$$

from which we can conclude that S *diverges*.

12.3. INFINITE SUMS

Interestingly, it can be shown that sum (12.53) will diverge even if we remove all terms whose denominators are *not* prime numbers. This is quite unexpected, given that the density of prime numbers approaches zero as $n \to \infty$ (according to the Prime Number Theorem). On the other hand, if we were to remove all terms whose denominator contains the digit 9 (such as 1/9, 1/59, 1/901, etc.) the series would converge to a number that lies between 22.4 and 23.3.

Example 12.4. In contrast to the harmonic series, the *geometric series*

$$S = \sum_{k=0}^{\infty} aq^k \qquad (12.58)$$

(where a is an arbitrary constant) exhibits very predictable behavior. Legend has it that this sum made its first appearance at the court of an Indian king, who had invited the inventor of the game of chess in order to honor him for his discovery. When asked to name his reward, the inventor (who was apparently a mathematician) offered the following proposal: "One grain of wheat for the first square on the board, two for the second, four for the third, and so on until the sixty fourth square." The king was pleased by the mathematician's modesty until he realized that

$$S = \sum_{n=0}^{63} 2^n \qquad (12.59)$$

is actually a staggering amount (roughly 1.84×10^{19} grains). To get a sense for how large this number is, let us assume that each grain has a diameter of 1 millimeter (which is a very conservative estimate). In that case, a line composed of this many grains would be nearly two light years long!

The partial sum of the first n terms in the geometric series can be calculated by observing that

$$S_{n-1} = \sum_{k=0}^{n-1} aq^k = a + aq + aq^2 + \ldots + aq^{n-1} \qquad (12.60)$$

and

$$qS_{n-1} = \sum_{k=0}^{n-1} aq^{k+1} = aq + aq^2 + aq^3 + \ldots + aq^n \qquad (12.61)$$

The only two terms that these sums do *not* have in common are a and aq^n, which implies that

$$S_{n-1} - qS_{n-1} = (1-q)S_{n-1} = a - aq^n \qquad (12.62)$$

and therefore
$$S_{n-1} = \frac{a - aq^n}{1-q} = a \cdot \frac{1-q^n}{1-q} \tag{12.63}$$

When $-1 < q < 1$, this sum obviously converges to
$$S = \lim_{n \to \infty} S_{n-1} = a \cdot \frac{1}{1-q} \tag{12.64}$$

An interesting special case arises when we set $a = 0.9$ and $q = 0.1$ in expression (12.58). The corresponding sum has the form
$$S = \sum_{k=0}^{\infty} 0.9 \cdot (0.1)^k = 0.99999\ldots = 0.\bar{9} \tag{12.65}$$

with infinitely many nines in the decimal expansion. On the other hand, substituting $a = 0.9$ and $q = 0.1$ into (12.64) produces $S = 1$, so we can legitimately argue that $0.\bar{9}$ and 1 are actually the *same* number.

Even more interesting are the two borderline scenarios that correspond to $q = 1$ and $q = -1$, respectively. In the first case the geometric series obviously diverges, since
$$S_n = \sum_{k=0}^{n} aq^k = \sum_{k=0}^{n} a = a(n+1) \tag{12.66}$$

The second scenario is a bit more complicated, since the signs of the terms *alternate*. To get a better sense for what actually happens in this case, let us set $a = 1$ and examine the sum
$$S = \sum_{k=0}^{\infty} (-1)^k = 1 - 1 + 1 - 1 + 1 - \ldots \tag{12.67}$$

If we group the terms as
$$S = (1-1) + (1-1) + (1-1) + \ldots \tag{12.68}$$

we obtain $S = 0$, while arranging them as
$$S = 1 + (-1+1) + (-1+1) + (-1+1) + \ldots \tag{12.69}$$

produces $S = 1$. A third possibility would be to express S as
$$S = 1 - (1 - 1 + 1 - 1 + 1 - \ldots) = 1 - S \tag{12.70}$$

which yields
$$2S = 1 \tag{12.71}$$

and therefore $S = 1/2$.

12.3. INFINITE SUMS

How should we interpret this strange result? Since the infinite sum gives different answers depending on how we arrange the terms, we must conclude that it *doesn't converge*. Note, however, that in this case divergence does not imply that the result necessarily blows up to infinity - it simply means that it is non-unique. The following quote by Niels Henrik Abel illustrates the frustration that mathematicians have often experienced in working with such sums:

> "Divergent series are the invention of the devil. ... By using them, one may draw any conclusion he pleases and that is why these series have produced so many fallacies and so many paradoxes." [6]

Example 12.5. An interesting (and counterintuitive) property of infinite sums has to do with the fact that the result sometimes depends on the *order in which the terms are added up*. This is clearly not the case with finite sums, whose value remains unchanged when we rearrange the terms. Indeed, it is easy to see that the sum of, say, the first five natural number equals 15 regardless of how we order them:

$$1 + 2 + 3 + 4 + 5 = 1 + 3 + 5 + 2 + 4 = 5 + 4 + 1 + 2 + 3 = 15 \quad (12.72)$$

To see why this sort of invariance is *not* guaranteed when the number of terms is infinite, consider the sum

$$S = \sum_{n=1}^{\infty} \frac{(-1)^{n+1}}{n} = 1 - 1/2 + 1/3 - 1/4 + \ldots \quad (12.73)$$

which is a variant of the harmonic series where the coefficients have alternating signs. It is not difficult show that this sum converges, and that its limit is $\ln 2$. In order to do that, we should first observe that

$$\sum_{k=0}^{\infty} y^k = \frac{1}{1-y} \quad (12.74)$$

(which follows directly from (12.64) by setting $a = 1$ and $q = y$). If we now integrate both sides of (12.74) from 0 to x, we obtain

$$\sum_{k=0}^{\infty} \int_0^x y^k dy = \sum_{k=0}^{\infty} \frac{x^{k+1}}{k+1} = \int_0^x \frac{1}{1-y} dy = -\ln(1-x) \quad (12.75)$$

When we substitute $x = -1$ into this expression, it takes the form

$$\sum_{k=0}^{\infty} \frac{(-1)^{k+1}}{k+1} = -\ln 2 \quad (12.76)$$

Setting $n = k + 1$ and multiplying both sides by -1, (9.75) finally becomes

$$\sum_{n=1}^{\infty} \frac{(-1)^{n+1}}{n} = \ln 2 \qquad (12.77)$$

which is what we set out to prove.

What makes this particular sum intriguing is the fact that its value changes if we modify the order of summation. To see how something like this is possible, let us multiply the sum in (12.73) by $1/2$, which produces

$$S/2 = 1/2 - 1/4 + 1/6 - 1/8 + \ldots \qquad (12.78)$$

If we now align the two sequences in the manner shown in Table 12.6 and add up the matching terms, we obtain

$$\bar{S} = 1 + 1/3 - 1/2 + 1/5 + 1/7 - 1/4 + \ldots \qquad (12.79)$$

It is not difficult to see that $\bar{S}$ has exactly the same terms as S, only added in a different order. One might therefore expect that $\bar{S} = \ln 2$, but this is clearly not the case, since $\bar{S} = S + S/2 = (3/2) \ln 2$.

$S:$	1	$-1/2$	$1/3$	$-1/4$	$1/5$	$-1/6$	$1/7$	$-1/8$	$\cdots$
$S/2:$		$1/2$		$-1/4$		$1/6$		$-1/8$	$\cdots$
		$\downarrow$		$\downarrow$		$\downarrow$		$\downarrow$	
$\bar{S}:$	1		$1/3$	$-1/2$	$1/5$		$1/7$	$-1/4$	$\cdots$

Table 12.6. Sum of sequences S and $S/2$.

Infinite sums of this sort are not uncommon, and are known as *conditionally convergent series*. More precisely, we say that an infinite sum

$$S = \sum_{n=1}^{\infty} a_n \qquad (12.80)$$

is conditionally convergent if the limit

$$\lim_{m \to \infty} \sum_{n=1}^{m} a_n \qquad (12.81)$$

exists and is *finite*, while

$$\lim_{m \to \infty} \sum_{n=1}^{m} |a_n| \qquad (12.82)$$

does *not*. Riemann proved that a conditionally convergent series can actually be rearranged to converge to *any value at all* (this surprising result is known as the Riemann Series Theorem).

Chapter 13

Introduction to Infinite Sets

13.1 What is a Set?

"A set is a Many that allows itself to be thought of as a One" Georg Cantor [7]

13.1.1 The Intuitive Approach

The metaphor that is most commonly used to describe a set is that of a "container" into which other objects are placed. These objects can be pretty much anything, including other containers. The schematic diagram in Fig. 13.1 shows how this "mental image" allows us to distinguish between the empty set ∅ (which can be viewed as an empty container), and a set like {∅} (which corresponds to a container that includes another empty container). It also helps us visualize sets such as {∅, {a}, {b, c}} whose interpretation is provided in Fig. 13.2.

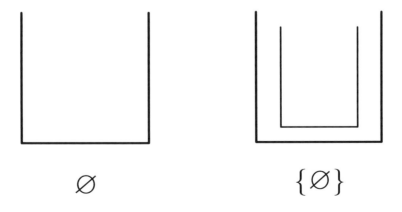

Figure 13.1: Illustration of the difference between ∅ and {∅}.

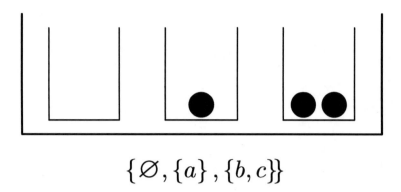

$\{\emptyset, \{a\}, \{b, c\}\}$

Figure 13.2: Illustration of a set whose members are other sets.

Although the "container metaphor" is undoubtedly useful and intuitively appealing, we shouldn't lose sight of the fact that it imposes certain restrictions on the kinds of sets that we can conceive of. Perhaps the most obvious example of how this metaphor can fail are sets that include themselves as members. One could, of course, dismiss such an idea on the grounds that it appears to have no practical significance, but such arguments are not particularly persuasive. Indeed, the history of mathematics is full of examples where seemingly "useless" intellectual constructs eventually led to important discoveries, so we have no a priori reason to believe that studying "unconventional" sets and their properties will necessarily be a futile effort.

If we agree with this premise, we then need to examine how such mathematical objects can be described in a consistent manner. One way to do that would be to think of sets as *directed graphs*. This new "conceptual metaphor" allows us to represent sets as vertices in a graph whose edges indicate how these sets are related. The schematic diagrams shown in Figs. 13.3 and 13.4 illustrate how this approach works in practice. In Fig. 13.3, the direction of the edge implies that set B is a member of set A (since the "arrow" points from node A to node B). By the same logic, the graph in Fig. 13.4 indicates that sets A and B are members of set C, which, in turn, is a member of set D.

In this representation, sets that contain themselves as members correspond to vertices with *self-loops*. Graph theory allows for such a possibility, but it remains unclear what happens when these unusual mathematical objects (which are referred to as *hypersets*) are incorporated into set theory. One of the immediate consequences is that any set X can now satisfy *either* $X \in X$ or $X \notin X$, which was not possible in the model that was based on the "container metaphor".

To see why this is potentially problematic, let us consider a set R which consists of all sets such that $X \notin X$. Such a definition gives rise to two possible

13.1. WHAT IS A SET?

scenarios:

(a) If we assume that $R \in R$ it follows that this set includes itself as a member. However, by definition R can contain only sets that do *not* include themselves as a member, which leads to a contradiction.

(b) If we assume that $R \notin R$, this means that R is *not* be a member of itself. However, all such sets must belong to R, so it follows that $R \in R$. We now have a situation where $R \notin R$ implies $R \in R$, which is once again a contradiction.

Figure 13.3: Graph-theoretic representation of $B \in A$.

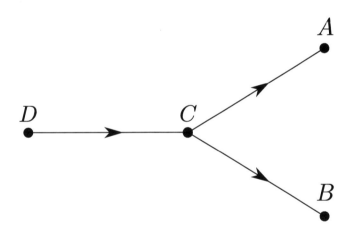

Figure 13.4: Graph-theoretic representation of $A \in C$, $B \in C$ and $C \in D$.

Since neither of these two possibilities are logically acceptable, we obviously have a paradox (which was first discovered by Bertrand Russell, and therefore bears his name). One way to resolve it would be to conclude that R *is not a set*, in which case the logical difficulties that we encountered above are automatically eliminated. But if R is not a set, then what is it?

Mathematicians refer to any collection of objects as a *class*, which may or may not be unifiable into a set. In the latter case, we say that the class is *proper*, and R would seem to fit into this category. If we adopt Cantor's intuitive definition of a set (which was cited at the beginning of this chapter), what we really have here is a "many" that can *not* be thought of as a "one". This sounds almost mystical and stretches the imagination beyond its natural limits, so it should come as no surprise that set theory has aroused the interest of not just mathematicians, but also philosophers, psychologists and even theologians.

13.1.2 The Axiomatic Approach

In order to avoid the potential difficulties that arise when we attempt to define sets intuitively, mathematicians have devised an alternative approach that allows us to introduce the notion of a set without resorting to conceptual metaphors. This approach is purely formal, and defines a set as any mathematical object that satisfies a collection of nine axioms known as the *Zermelo – Fraenkel* (ZF) *axioms*. There is nothing in these axioms that requires us to envision sets as containers, graphs, or any other entity related to our everyday experience - they are completely independent and self-contained. Another advantage of the ZF model is that it effectively eliminates certain problematic concepts from set theory (such as sets that are members of themselves, for example). As a result, Russell's paradox cannot arise, and neither can some other logical conundrums that plagued earlier versions of set theory.

Perhaps the most interesting among the ZF axioms is the Axiom of Choice, which asserts that any set can be well-ordered (this effectively means that it has a first element, and that there is a precise way to determine the "successor" of each member of the set). It can be shown that this axiom is consistent with the other ZF axioms, but the same holds true for its negation as well. As a result, many mathematicians consider it to be controversial, and view it as "less fundamental" than the other eight. This outlook is so prevalent that a distinction is usually made in the literature between proofs that explicitly utilize the Axiom of Choice, and those that do not.

Although it was designed to remove all ambiguities from set theory, the axiomatic approach has its own share of "imperfections". One of them stems from the fact that the ZF axioms allow for *multiple interpretations* of basic set theoretic concepts (including the notion of a set itself). This is a direct consequence of the Löwenheim-Skolem theorem, which addresses the following fundamental question: Given the nine ZF axioms and the theorems that can be derived from them, can we precisely understand what terms like "set" and "member" really mean without any prior exposure to these concepts?[8] Can we tell, in other words, what a set *is* by simply listing all the axioms and theorems of set theory? The Löwenheim-Skolem theorem shows that we can *not*, and tells us that the mathematical objects that are described by the ZF theory could very well be interpreted as *natural numbers* (which, as we shall see, are sets in their own right, but are only a miniscule portion of all possible sets).

Since the ZF model does not allow us to formally distinguish sets from natural numbers, it is reasonable to expect that the knowledge derived from

13.1. WHAT IS A SET?

its theorems might have certain inherent limitations. One could legitimately wonder, for example, whether this theory can tell us anything meaningful about sets that do *not* correspond to natural numbers, given that its "conceptual vocabulary" has no need for them (the Löwenheim-Skolem theorem implies that set theory would remain unchanged even if natural numbers were the *only* type of sets that existed).

The questions raised above have serious epistemological implications, and suggest that the foundations of modern set theory may be shakier than we would like to admit. One could, of course, attempt to diffuse the problem by arguing that our current understanding of set theory is incomplete, and that some of the difficulties that are associated with the interpretation of the Löwenheim-Skolem theorem will eventually be resolved once better models are developed. Such expectations are not unreasonable, since it is entirely possible that some of our fundamental assumptions about sets could, in fact, change. The difficulty with such arguments is that we cannot properly speculate about the scope of a theory without "stepping outside" it. It may indeed be the case that our understanding of set theory is inherently limited, but how could we possibly know that? All that we can reliably say at the present time is that the current version of set theory is not entirely complete, and nothing more.

A pragmatic approach to this problem might be to argue that we never really think of concepts such as "set" and "member" in purely formal terms. We all have an intuitive understanding of what a set is, which is acquired through experience long *before* we are actually exposed to set theory. This means that we already have a "context" for interpreting the axioms and theorems of ZF theory when they are presented to us, and are therefore unlikely to misunderstand what they mean (although this is theoretically possible, according to the Löwenheim-Skolem theorem). We can therefore conclude that the problem posed by this theorem is a purely technical one, and has no practical effect on the way mathematicians go about their work. Indeed, I seriously doubt that anyone bases their understanding of what a set is on axioms and formal propositions alone - we also use our intuition and experience, and that is what ultimately "steers" us toward the intended interpretation.

Gödel's First Incompleteness Theorem is another result that places certain constraints on the Zermelo-Fraenkel axiomatic model.[9] This theorem tells us that any internally consistent formal system with a finite set of axioms will always include a statement S that can neither be proved nor disproved from "within". Gödel showed that the problem cannot be resolved by changing the axioms or adding new ones. Doing so could perhaps allow us to determine whether or not S is true, but there would always be some other statement S^* that is unprovable in the formal system defined by the new axioms. There is, in other words, no way to classify *all* the propositions in such a system as either true or false - some will necessarily be *undecidable*.

Although Gödel's theorem has its origins in formal logic and arithmetic, it applies to set theory as well. This is because natural numbers and the basic operations that pertain to them can always be equivalently represented in set-theoretic terms (in the manner outlined in Section 13.2). In view of that, it is reasonable to expect that there will be statements about sets that cannot be proved or disproved within the axiomatic framework of ZF theory.

13.1.3 Pure Sets

The "container metaphor" introduced at the beginning of this chapter suggests that sets can be understood as *collections* of elements. These elements can be pretty much anything that we can think of, the only requirement being that they must exist *before* they are assembled into a "unified whole". Since such a definition is extremely broad, mathematicians usually prefer to study specific types of sets whose properties are "representative" of all other possibilities. One such class are so-called *pure sets*, whose members can only be other sets. The term "pure" underscores the fact that such sets are independent of the physical world, and would presumably continue to exist even in a universe that is completely devoid of matter and energy (and therefore of mathematicians as well).

What do pure sets look like? We can think of them as a class of mathematical objects that is formed *iteratively*, starting from the empty set. In each step of this process, all sets from the previous stage are combined into *new* sets, producing an ever increasing collection. The following example illustrates how the first four stages work.

Stage 1. The only set in this stage is the empty set, $\emptyset$.

Stage 2. In the second step we have two choices - we can form a set that includes $\emptyset$, and a set that does not. In the first case, we obtain set $S_1 = \{\emptyset\}$, and in the second $S_0 = \emptyset$.

Stage 3. Since the previous stage produced two sets, for any new set that is constructed in Stage 3 we need to determine whether S_0 and S_1 should be included. This is equivalent to answering two Yes/No questions, which give rise to four different possibilities:

$$\begin{aligned}
\text{No, No} &\longleftrightarrow S_{00} = \emptyset \\
\text{No, Yes} &\longleftrightarrow S_{01} = \{S_1\} = \{\{\emptyset\}\} \\
\text{Yes, No} &\longleftrightarrow S_{10} = \{S_0\} = \{\emptyset\} \\
\text{Yes, Yes} &\longleftrightarrow S_{11} = \{S_0, S_1\} = \{\emptyset, \{\emptyset\}\}
\end{aligned}$$

Stage 4. In this stage, new sets are formed from S_{00}, S_{01}, S_{10} and S_{11}, which means that we have to ask *four* Yes/No questions. This give rise to $2^4 =$

13.1. WHAT IS A SET?

16 different possibilities, which are described below (to simplify the notation, in this list we denoted Yes with subscript 1 and No with 0).

$$S_{0000} = \emptyset$$
$$S_{0001} = \{S_{11}\} = \{\{\emptyset, \{\emptyset\}\}\}$$
$$\vdots$$
$$S_{1111} = \{S_{00}, S_{01}, S_{10}, S_{11}\} = \{\emptyset, \{\{\emptyset\}\}, \{\emptyset\}, \{\emptyset, \{\emptyset\}\}\}$$

The pattern that emerges from Stages 1 - 4 suggests that each step of this process produces all the sets that were constructed before, as well as a number of new ones. One can see this very clearly from the list that corresponds to Stage 4 - the sets that already existed prior to this step are $S_{0000} = \emptyset$, $S_{1000} = \{S_{00}\} = \{\emptyset\}$, $S_{0010} = \{S_{10}\} = \{\{\emptyset\}\}$ and $S_{1010} = \{S_{00}, S_{10}\} = \{\emptyset, \{\emptyset\}\}$, while the remaining 12 sets appear for the first time.

To get a sense for how quickly the number of new sets increases, suppose that Stage n of this process produces k different sets. In that case, Stage $n+1$ will contain 2^k sets, each of which corresponds to a different pattern of k Yes/No answers. In the following step, we obtain 2^{2^k} sets, which is a much larger number than 2^k. Just how much larger 2^{2^k} can be is illustrated by observing that Stage 4 corresponds to $k = 16$, while Stage 5 produces $2^{16} = 65,536$ sets. In Stage 6 this number grow to $2^{2^{16}} = 2^{65,536}$ sets, and by Stage 7 we come to a point where the number is so large that it becomes difficult to imagine.

In order to symbolically represent such enormous numbers, mathematicians have devised a special kind of notation whose general pattern is illustrated in Table 13.1. In this scheme, the symbol $^n2$ represents 2 *tetrated* to n, which allows us to express the number of sets obtained in Stage 7 in compact form as ^{52}.

Number of sets in Stage 2	$\to$	$2^1 = 2^{2^0} \equiv {}^02$
Number of sets in Stage 3	$\to$	$2^2 = 2^{2^1} \equiv {}^12$
Number of sets in Stage 4	$\to$	$2^4 = 2^{2^2} \equiv {}^22$
Number of sets in Stage 5	$\to$	$2^{16} = 2^{2^{2^2}} \equiv {}^32$
$\vdots$		$\vdots$

Table 13.1. The process of "tetration".

13.2 Ordinals

The notion that sets can be constructed iteratively has several important implications, one of which is that we can think of *numbers as sets*. In order to see what this means, we should first note that in our everyday experience, numbers are typically used in one of two different ways - they either tell us the *position* of an element in an ordered set, or they describe the *size* of the set. In the first case, we refer to them as *ordinals*, while in the second case we call them *cardinals*.

In the domain of finite numbers, cardinals and ordinals are related in a very straightforward way. As an illustration, consider the set $S = \{1, 2, 5, 7\}$ whose cardinality is defined by the total number of elements that it contains (which is obviously 4). If we choose to think of the elements of S as a sequence: $x_1 = 1$, $x_2 = 2$, $x_3 = 5$, $x_4 = 7$, the order becomes important, and we can identify the *last one* to be labeled (or counted). The ordinal number for this element is 4, and it is easily seen that it matches the cardinal number of the set as a whole. [10]

In order to properly define ordinal numbers, it is first necessary to describe how "order" can be established in a set. We will say that a set is *well ordered* if it satisfies the following two conditions:

Condition 1. An element must be defined as the *first* member of the set.

Condition 2. Each element in the set has a *successor*, which is uniquely defined by the ordering scheme.

Conditions 1 and 2 are quite flexible, and allow us to impose order in many different ways. Consider, for example, how this can be done for a set that consists of 5 people. We can obviously organize its members by their height, weight, age, income, etc., and each of these criteria would produce a different ordering. The *structure* of these orderings, however, is the same in all cases - there is always a first element, and a uniquely defined successor.

Once a set is well ordered, each of its members has a definite *position* in the overall hierarchy. Numbers that correspond to these positions are referred to as *ordinals*, and we can use them to index the elements as $\{x_1, x_2, \ldots, x_n\}$. In such a set, we will say that $x_i < x_j$ if element x_j appears *after* element x_i in the set. When used in this context, the notion of "greater than" obviously refers to the location of elements in the set (and not necessarily their "size").

The connection between ordinals and sets becomes apparent if we observe that every ordinal is completely defined by the *set of its predecessors* (since the ordering scheme uniquely determines the successor of the elements that are already in place). This property allows us to treat the two concepts as *equivalent*, and identify any finite ordinal k with a well ordered set of the form $\{0, 1, 2, \ldots, k-1\}$.

13.2. ORDINALS

We can generalize this idea even further, by replacing numbers with *pure sets*. To get a sense for how this works, we should first recall that pure sets are constructed in stages, which can be numbered. This feature allows us to identify a "representative set" in each stage, and associate it with the corresponding ordinal. Since the construction process begins with the empty set, it makes sense to associate $S_0 = \emptyset$ with 0, and then proceed in the manner indicated in Table 13.2. Note that in each step of this process the elements of the next set are sets that have *already been formed* up to that point. This is consistent with our earlier assumption that each ordinal is defined by the set of predecessors.

If we compare Table 13.2 to the construction process described in Section 13.1.3, it is not difficult to see that set S_k formed in this manner appears for the first time in stage $k+1$ of the iterative process. Indeed, S_2 is identical to S_{11} (which appears for the first time in Stage 3 of this process), and that S_3 matches $S_{1011} = \{S_{00}, S_{10}, S_{11}\}$ (whose first appearance is in Stage 4).

Stage 1	$S_0 = \emptyset$	$\longleftrightarrow$ 0
Stage 2	$S_1 = \{S_0\} = \{\emptyset\}$	$\longleftrightarrow$ 1
Stage 3	$S_2 = \{S_0, S_1\} = \{\emptyset, \{\emptyset\}\}$	$\longleftrightarrow$ 2
Stage 4	$S_3 = \{S_0, S_1, S_2\} = \{\emptyset, \{\emptyset\}, \{\emptyset, \{\emptyset\}\}\}$	$\longleftrightarrow$ 3
$\vdots$	$\vdots$	$\vdots$
Stage $n+1$	$S_n = \{S_0, S_1, \ldots, S_{n-1}\} = \{\emptyset, \{\emptyset\}, \ldots\}$	$\longleftrightarrow$ n
$\vdots$	$\vdots$	$\vdots$

Table 13.2. Correspondence between numbers and pure sets.

Given that sets can be represented as graphs, it is reasonable to expect that one could define numbers in this way as well. This is indeed possible, if we envision zero (and therefore the empty set as well) as a single node, and describe any number $N > 0$ in terms of a directed graph in which each node has exactly N neighbors. In this representation, the nodes would be connected hierarchically, in the sense that one node (the so-called "root node") would have N outgoing edges, the next one would have $N - 1$ outgoing edges and so on until the final node, which has no outgoing edges. Figs. 13.5 - 13.7 show how numbers 0, 1, 2 and 3 would be represented by such graphs, with the set theoretic model acting as an "intermediary" (note that in each case the "root node" corresponds to the number).

13.2.1 Infinite Ordinals

The equivalence between numbers and pure sets is useful because it allows us to introduce the notion of *infinite ordinals*. Our ability to do so relies on two fundamental assumptions:

Assumption 1. Given any ordinal a, there is a "smallest" ordinal greater than a (which is denoted $a + 1$). This number is the *successor* of a in the list.

Assumption 2. If $A = \{a_n\}$ is an ordered set of ordinals such that $a_0 < a_1 < \ldots < a_k < \ldots$ then there exists a "smallest" ordinal which is greater than all a_n.

These two assumptions are trivial when it comes to finite ordinals, and require no additional explanation. What is less obvious, however, is the fact that they can be extended in a way that allows us define an entire hierarchy of *infinite* ordinals. This requires a "cognitive metaphor" which implicitly assumes that an endless iterative process has an *actual result*, and that this result is *unique* (in the introduction to Chapter 12, we referred to it as the Basic Metaphor of Infinity).

If we use such a metaphor to generalize Assumptions 1 and 2, we can extend the construction process described in Table 13.2 beyond finite numbers, and can define the *first infinite ordinal* as [11]

$$\omega \equiv S_\omega = \{S_0, S_1, \ldots, S_n, \ldots\} = \{\emptyset, \{\emptyset\}, \{\emptyset, \{\emptyset\}\}, \ldots\} \qquad (13.1)$$

Such a set is well defined, since all the elements that make it up have already been formed at that point.

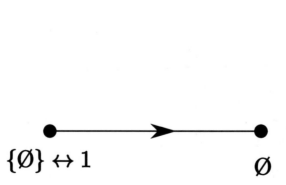

Figure 13.5: Graph-theoretic representation of numbers 0 and 1.

13.2. ORDINALS

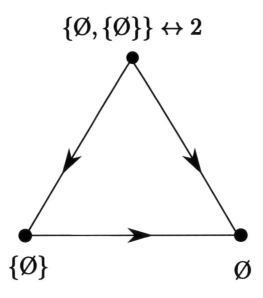

Figure 13.6: Graph-theoretic representation of number 2.

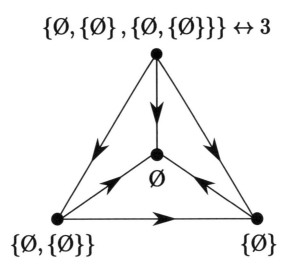

Figure 13.7: Graph-theoretic representation of number 3.

The procedure described in Table 13.2 shows how ω can be constructed from finite sets, but it does not tell us what this ordinal represents. Since it corresponds to an infinite set, it is clear that ω itself is *not* a natural number, and is a mathematical entity of a very different kind. One could perhaps think of it as the set of natural numbers viewed as a "whole", which is something that can be defined but not visualized. However we choose to interpret it, the fact that ω is an ordinal implies that it must have a successor, which is denoted by $\omega+1$ (this follows directly from Assumption 1). Note that in this case, the "+" sign does not represent conventional addition, and serves instead to indicate the position of $\omega + 1$ in the overall hierarchy of ordinals.

This process can obviously be continued by defining $\omega+2$ as the immediate successor of $\omega+1$, $\omega+3$ as the immediate successor of $\omega+2$, and so on (the first three infinite ordinals constructed in this manner are shown in Table 13.3). As we continue to add elements $\omega + 4, \omega + 5, \ldots, \omega + k, \ldots$ to this sequence, we will eventually come to a point when a second conceptual metaphor becomes necessary. This occurs when $k \to \infty$, and the resulting number can be represented as $\omega \cdot 2$ (such number can be viewed as the "successor" of all ordinals of the form $\omega + n$, and its existence is ensured by the Axiom of Replacement in the ZF model).

$$\omega + 1 \equiv S_{\omega+1} = \{S_0, S_1, \ldots, S_n, \ldots, S_\omega\}$$
$$\omega + 2 \equiv S_{\omega+2} = \{S_0, S_1, \ldots, S_n, \ldots, S_\omega, S_{\omega+1}\}$$
$$\omega + 3 \equiv S_{\omega+3} = \{S_0, S_1, \ldots, S_n, \ldots, S_\omega, S_{\omega+1}, S_{\omega+2}\}$$
$$\vdots$$

Table 13.3. Construction of infinite ordinals.

It is interesting to note in this context that commutativity (which implies that $a+b = b+a$) ceases to hold when one of the terms in the sum is an *infinite* ordinal. To see this a bit more clearly, let us compare ordinals $a = \omega + 1$ and $b = 1+\omega$. The first number is simply the successor of ω, while b represents the successor of all numbers of the form $1+n$ (which happens to be ω, since all of these numbers are finite). As a result, we can conclude that

$$1 + \omega = \omega \neq \omega + 1 \qquad (13.2)$$

Something similar holds true for multiplication as well. The number $c = \omega \cdot 2$, for example, is equivalent to the successor of all ordinals of the form $\omega+n$, and is therefore an infinite ordinal that is *distinct* from ω itself. The number $d = 2 \cdot \omega$, on the other hand, has a very different meaning, and can be interpreted as the successor of all numbers of the form $2n$. Since these numbers are finite, it follows that

$$2 \cdot \omega = \omega \neq \omega \cdot 2 \qquad (13.3)$$

How far can the iterative procedure described in Table 13.3 continue? We can obviously define the successors of $\omega \cdot 2$ as $\omega \cdot 2 + 1$, $\omega \cdot 2 + 2$, ... and introduce the ordinal $\omega \cdot 3$ as the successor of all numbers of the form $\omega \cdot 2 + n$. This will eventually lead us to ordinals such as $\omega \cdot 4$, $\omega \cdot 5$, ... and then to ω^2, ω^3, ... The successor of all ordinals of the form ω^n is denoted by ω^ω (or alternatively as $^2\omega$) which is followed by $^3\omega \equiv \omega^{\omega^\omega}$, $^4\omega$, ... , $^n\omega$, and so on.

Is there anything else beyond numbers of the form $^n\omega$? Assumption 1 suggests that there must be, and we define the first such ordinal as $\varepsilon_0 = {}^\omega\omega$. Such a number is impossible to imagine, but this does not imply that it is any way "artificial". If we adopt the "realist" position, and assume that mathematical forms have an objective existence regardless of our ability to grasp them, then ε_0 is a perfectly legitimate member of the hierarchy. We can obviously continue this process beyond ε_0, forming ordinals such as $\varepsilon_0 + 1$, $\varepsilon_0 + 2$, and so on.

The story does not end there, however, since we can now ask whether *all* ordinals formed in this manner can be collected into a single set O_n. Let us assume for a moment that this is possible. In that case, Assumption 2 would imply that there must be an ordinal Ω that *succeeds* all the members of this set. Being an ordinal, however, this number would necessarily belong to O_n, implying that $\Omega < \Omega$ (which is obviously a contradiction, since an ordinal cannot succeed itself). We therefore have something that looks very much like a paradox - it seems that no well ordered set is "large enough" to contain all the ordinals.

How should we interpret mathematical entities such as O_n and Ω? This is the point where the line between mathematics and metaphysics becomes blurred. If sets are understood as entities that can be known in an objective way and can be manipulated by the human mind, then O_n and Ω obviously do not fall into this category. We may have some idea of what they are and talk about them in a more or less meaningful way, but they ultimately remain beyond our reach.

13.3 Cardinals

We noted earlier that cardinals represent a measure for the *size* of a set. In the domain of finite sets, sizes can be compared very easily by counting the number of elements. This approach cannot be extended to infinite sets, however, since the traditional notion of counting does not apply any more. In order to resolve this difficulty, Cantor introduced a different method for comparing two sets, which is based on the notion of *pairability*. More specifically, he proposed that two sets have the same size (or *cardinality*) if and only if there is a 1-to-1 correspondence between their elements.

From a mathematical perspective, Cantor's idea is perfectly coherent, since it applies equally well to finite and infinite sets. However, some of the results that it leads to are highly counterintuitive and defy our traditional notion of "size". In the following, we will see a number of examples that illustrate exactly how different infinite sets are from their finite counterparts.

Example 13.1. The simplest infinite set that we can imagine is the set of natural numbers, $N = \{0, 1, 2, \ldots\}$. The cardinality of this set is denoted by $\aleph_0$, which is pronounced *aleph zero* (aleph is the first letter of the Hebrew alphabet). When we compare this set to the set of all rational numbers, Q, it seems "obvious" that Q is the larger of the two, since it contains all natural numbers as well as all the fractions that lie between them. It is not difficult to show, however, that the two sets actually have the *same* cardinality. According to Cantor's comparison method, all that we need to do is show that a 1-to-1 mapping exists between their elements.

The easiest way to construct such a mapping is to arrange all rational numbers in the manner shown in Table 13.4, and observe that we can systematically construct a "path" that connects them.

$$
\begin{array}{ccccc}
1/1 \rightarrow & 2/1 & 3/1 \rightarrow & 4/1 & 5/1 \cdots \\
\swarrow & \nearrow & \swarrow & \nearrow & \\
1/2 & 2/2 & 3/2 & 4/2 & 5/2 \\
\downarrow \nearrow & & \swarrow & \nearrow & \\
1/3 & 2/3 & 3/3 & 4/3 & 5/3 \\
& \swarrow & \nearrow & & \\
1/4 & 2/4 & 3/4 & 4/4 & 5/4 \\
\downarrow \nearrow & & & & \\
1/5 & 2/5 & 3/5 & 4/5 & 5/5 \\
\vdots & & & & \ddots \\
\end{array}
$$

Table 13.4. Scheme for "enumerating" rational numbers.

The construction process can be described as a sequence of four steps which are repeated iteratively:

1) One step to the *right*.

2) All the way *down diagonally*.

3) One step *down*.

4) All the way *up diagonally*.

13.3. CARDINALS

It is not difficult to see that by following this scheme we can visit *all* the elements in the two dimensional array, and number them along the way. In doing so, we will never run out of natural numbers (which helps explain why the two sets can be thought of as "equal" in size).[12]

Example 13.2. If the set of all rational numbers is no larger than N, what (if anything) is? In this example, we will show that the set of real numbers on the interval [0 1] falls into this category, since its cardinality is *greater* than $\aleph_0$. The proof that follows was proposed by Cantor, and is considered to be one of the most elegant demonstrations in modern mathematics.

Cantor approached this problem by assuming that the *opposite* is true, and that we can form an infinite list in which each real number $r(n)$ is uniquely matched with a positive integer n. In that case, our list could look something like the one shown in Table 13.5.

$$r(1): \quad 0.\mathbf{5}3210\ldots$$
$$r(2): \quad 0.8\mathbf{6}343\ldots$$
$$r(3): \quad 0.75\mathbf{3}12\ldots$$
$$r(4): \quad 0.099\mathbf{2}0\ldots$$
$$r(5): \quad 0.1348\mathbf{0}\ldots$$
$$\vdots \qquad \vdots$$

Table 13.5. A hypothetical ordered list of real numbers

The idea now is to form a number that is *not* a member of the list. If we succeed, this will obviously contradict our initial assumption that the interval [0 1] has the same cardinality as N. To construct such a number, let us subtract 1 from the diagonal (boldfaced) entry in each $r(i)$ ($i = 1, 2, \ldots$), with the understanding that subtracting 1 from 0 produces 9. The number obtained by assembling the modified diagonal terms would then be

$$x = 0.45219\ldots \tag{13.4}$$

By construction, x differs from $r(1)$ in the *first* digit, from $r(2)$ in the *second*, from $r(3)$ in the *third*, and so on. In other words, x is different from *every single number* in the list, and is therefore not a part of it. This leads us to conclude that there is at least one number in [0 1] that cannot be paired with a natural number.

Example 13.3. Given that interval [0 1] belongs to the set of all real numbers, R, one would intuitively expect that its cardinality is smaller than that of R. It turns out, however, that this is *not* the case, and that every element of R can be uniquely mapped into a point on interval [0 1]. The schematic diagram in Fig. 13.8 shows what such a mapping might look like. It is not difficult to see that the lines originating in points a, b and c intersect the circle in different places. These intersection points have different projections onto [0 1], and can therefore be paired with distinct elements of this set. Since this holds true for *any* element of R, we can conclude that [0 1] and R have the *same* cardinality (this number is usually denoted by c).[13]

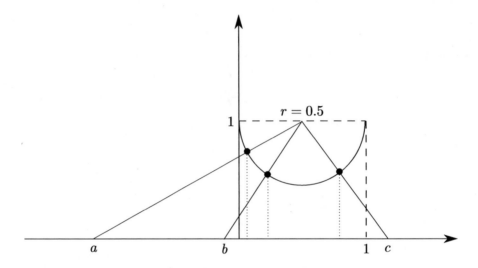

Figure 13.8: Mapping set R into interval [0 1].

13.3.1 Hierarchies of Infinities

Are there any sets whose cardinality is larger than c? One of Cantor's most striking discoveries was that there is actually an unlimited number of such sets, and that they can be arranged into a hierarchy. This result is based on the notion of a *power set*, which is introduced in the following example.

Example 13.4. Let us consider set $S = \{1, 2, 3\}$, and examine how many different subsets we can we form from its elements. In this case, the set of all possible subsets S has the form

$$P(S) = \{\emptyset, \{1\}, \{2\}, \{3\}, \{1,2\}, \{1,3\}, \{2,3\}, \{1,2,3\}\} \qquad (13.5)$$

and we refer to $P(S)$ as the *power set* of S. Since $P(S)$ is not a large set, we can easily verify that it contains 8 elements by simply counting them. There is

13.3. CARDINALS

a more elegant way to arrive at this result, however, which allows us to extend this approach to sets of any size. The basic idea is to recognize that forming a subset amounts to a sequence of binary choices. Indeed, for each element of S, we have the option of including it in the subset or not, which is essentially a Yes/No question. We can therefore think of $\{1,3\}$ as the subset that arises when we answer {Yes, No, Yes}, and $\emptyset$ as the subset that arises when we answer {No, No, No}. In general, for a set with k elements the number of different choices is 2^k, so we can conclude that the power set of S must contain $2^3 = 8$ elements.

Cantor recognized that this idea could be extended to infinite sets as well, and showed that the power set of any infinite set S is *larger* than S itself. The following theorem demonstrates how he proved this for the set of natural numbers (the method that he used is a variant of his famous "diagonal argument").

Theorem 13.1. Let $N = \{0, 1, 2, \ldots\}$ be the set of natural numbers, and let $P(N)$ denote its power set. Then there exists at least one element in $P(N)$ that cannot be mapped into set N (which is equivalent to saying that $P(N) > N$).

Proof. Let us assume that the theorem is *incorrect*, in which case there *is* a 1-to-1 correspondence between the elements of $P(N)$ and N. That would allow us to enumerate *all* the elements of $P(N)$, and list them in the manner shown in Table 13.6 (this table also includes the combination of Yes/No answers that gives rise to each subset).

Subset		0	1	2	$\cdots$
$S(0)$	$\longleftrightarrow$	No	No	No	$\cdots$
$S(1)$	$\longleftrightarrow$	Yes	No	Yes	$\cdots$
$S(2)$	$\longleftrightarrow$	No	Yes	Yes	$\cdots$
$\vdots$					

Table 13.6. A hypothetical ordered list of subsets of N.

Consider now the subset of N that corresponds to the sequence of answers {Yes, Yes, No, ...}. This subset must be an element of $P(N)$ by definition, but it does *not* belong to the list, since it differs from $S(0)$ in the first answer, from $S(1)$ in the second, from $S(2)$ in the third, and so on. We have therefore identified an element of $P(N)$ that has no "counterpart" in N, which contradicts our initial assumption. **Q.E.D.**

The cardinality of $P(N)$ is usually denoted as $2^{\aleph_0}$, in keeping with the way we describe the size of finite power sets.[14] Cantor's construction method allows us to go beyond that number, however, and define an entire hierarchy of infinite cardinals (each of which is greater than all preceding ones). It is not difficult to see, for example, that the power set of R has cardinality $\aleph_2 \equiv 2^{\aleph_1}$, since forming any given subset involves $\aleph_1$ binary choices. It can be shown that this set contains certain elements which have no counterparts in R, which implies that $\aleph_2 \equiv 2^{\aleph_1} > \aleph_1$. We can proceed in this manner indefinitely, forming a sequence of infinities such that

$$\aleph_0 < \aleph_1 < \aleph_2 < \ldots \tag{13.6}$$

(Cantor proved that the inequality $2^{\aleph_k} > \aleph_k$ holds for *all* $k \geq 0$).

The Continuum Hypothesis

Does the sequence shown in (13.6) exhaust all cardinals, or are there numbers that lie "in the gaps", so to speak? This turns out to be an *unanswerable* question. Cantor assumed that there are *no* cardinals between $\aleph_0$ and $\aleph_1$, but Gödel and Cohen subsequently showed that this conjecture (which is known as the Continuum Hypothesis), cannot be proved or disproved within the framework of the nine ZF axioms. If the Continuum Hypothesis is adopted as an additional axiom, we obtain standard set theory, while adopting its opposite (which is equally legitimate) leads to what is known as non-Cantorian set theory. The distinction between the two is not unlike the one we encounter in the context of geometry, where the adoption (or rejection) of Euclid's fifth postulate produces two very different mathematical frameworks.

What made Cantor think that there might be sets that are larger than N but smaller than R? To see that, we need to examine what a hierarchy of power sets looks like when this idea is applied to *finite* sets, starting with some set S_1 that contains a single member. The resulting situation is shown in Table 13.7 (which bears a striking resemblance to the iterative process described in Section 13.1.3).

The data shown in Table 13.7 gives rise to two important observations about the hierarchy of power sets:

1. The number of elements grows very rapidly after the first few steps (their size increases as $^n 2$).

2. The collection of sets $\{S_1, S_2, \ldots, S_n, \ldots\}$ is only a small fraction of all finite sets, since there are *many* sets which are larger than S_n but smaller than S_{n+1} as n increases.

13.3. CARDINALS

Set	Number of Elements
S_1	1
$S_2 = P(S_1)$	$2^1 = 2 = {}^0 2$
$S_3 = P(S_2)$	$2^2 = 4 = {}^1 2$
$S_4 = P(S_3)$	$2^4 = 16 = {}^2 2$
$S_5 = P(S_4)$	$2^{16} = 65,536 = {}^3 2$
$S_6 = P(S_5)$	$2^{65,536} = {}^4 2$
$\vdots$	$\vdots$

Table 13.7. A hierarchy of power sets.

Considerations of this sort led Cantor to assume that something similar might apply to the hierarchy of infinite sets as well, and that there could be sets whose cardinality lies between $\aleph_0$ and c. He was unable to find any such sets, however, so he eventually formulated the famous Continuum Hypothesis, which claims that $2^{\aleph_0}$ is indeed the *first* cardinal that follows $\aleph_0$ (and can therefore be labeled as $\aleph_1$). In its generalized version, this hypothesis refers to the *entire* hierarchy of infinite sets

$$\Xi = \{\aleph_0, 2^{\aleph_0}, 2^{2^{\aleph_0}}, \ldots\} \tag{13.7}$$

and assumes that there are no sets whose cardinality lies between any two consecutive elements of Ξ.

When interpreting the "undecidability" of the Continuum Hypothesis, we should keep in mind that this result was obtained under the assumption that the nine ZF axioms provide a precise description of sets and their properties. But what if this is not the case? Might there be a more complete set of axioms which would allow for a definitive resolution to Cantor's dilemma? Much of the recent research in set theory has focused on this possibility, and several new axioms have been proposed to that end. What is particularly interesting about these axioms is that most of them allow for the existence of "large cardinals", which lie beyond the scope of ZF set theory. These "inaccessible" numbers (which cannot be constructed from numbers smaller than themselves) offer some intriguing possibilities, but they do not tell us whether the axioms that give rise to them are actually true. The axioms that have been proposed so far are *not* intuitive, so there are no obvious criteria for choosing between them.

An alternative approach would be to conclude that the problems posed by the Continuum Hypothesis are simply too complex to be solved by adding new axioms. Perhaps the nine ZF axioms do, in fact, provide the best possible description of sets that we can produce, but this description may nevertheless be somewhat imprecise. When you think about it, this is not such a strange idea at all - we already know that certain aspects of the physical world elude our concepts and categories, so why should we rule out the existence of similar restrictions in the domain of mathematics? It is, after all, entirely possible that there is an independent "mathematical reality" which exists outside of the human mind, and that we may never have complete access to all of its truths.

13.3.2 Transfinite Arithmetic with Cardinals

The hierarchy of infinite sets shown in (13.6) is reminiscent of natural numbers (which appear as indices of different cardinals). In view of that, it would be reasonable to expect that these numbers are subject to basic arithmetical operations such as addition and multiplication. What follows are several examples that illustrate how these operations are defined, and how they might be interpreted.

Example 13.5. In the domain of finite numbers, addition is normally associate with *unions of sets*. A simple algebraic operation such as $2+3=5$, for example, can be thought of as the union of two disjoint sets whose cardinalities are 2 and 3 respectively. The result of this operation is obviously a set whose cardinality is 5. What is less clear, however, is how one can extend this line of reasoning to infinite sets.

To see how that can be done, consider what an operation such as $\aleph_0 + 1$ might mean. We can envision it as the union of set N and a set such as $\{-1\}$, which has a single member. If we combine these two sets, it is not difficult to see that each of its elements can be paired with an element of N. It therefore follows that

$$\aleph_0 + 1 = \aleph_0 \tag{13.8}$$

which is very different from what we encounter when adding finite numbers.

Example 13.6. In this example we will consider how operations such as $\aleph_0 + \aleph_0$ and $\aleph_0^2$ can be interpreted. The first one is actually quite straightforward, if we recognize that the set of even and odd numbers both have cardinality $\aleph_0$. Given that their union represents the set of natural numbers (whose cardinality is $\aleph_0$), we can conclude that [15]

$$\aleph_0 + \aleph_0 = \aleph_0 \tag{13.9}$$

In order to get a sense for what operation $\aleph_0^2$ means, it is helpful to consider the matrix shown in Table 13.4. The rows and columns of this matrix can both be paired with the elements of N, and therefore represent sets of cardinality $\aleph_0$.

13.3. CARDINALS

In view of that, the collection of all the elements in the matrix can be thought of as a set whose cardinality is $\aleph_0^2$. On the other hand, we know that the union of all these elements is actually the the set of all rational numbers, Q, whose cardinality is $\aleph_0$ (see Example 13.1 for a proof of this fact). As a result, we can say that

$$\aleph_0^2 = \aleph_0 \tag{13.10}$$

Example 13.7. What can we say about operations such as $\aleph_0 \cdot c$ and c^2 which involve cardinals that are larger than $\aleph_0$? The first of these operations can be envisioned as the union of intervals $[0\ 1], [1\ 2], [2\ 3], \ldots$, each of which has cardinality c. Since there are as many such intervals as there are natural numbers, we can say that the set formed from their union has cardinality $\aleph_0 \cdot c$. Observing that this set is identical to the set of real numbers, it easily follows that

$$\aleph_0 \cdot c = c \tag{13.11}$$

Interpreting what c^2 means is a bit trickier, and requires a schematic diagram like the one shown in Fig. 13.9.

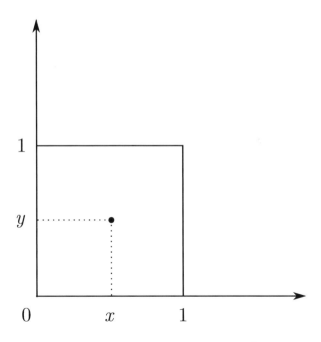

Figure 13.9: Demonstration that $c^2 = c$.

In this diagram, we will identify an arbitrary point in the unit square, whose coordinates can be described by a pair of real numbers (x, y). Since these numbers have the general form $x = 0.\,x_1\,x_2\,x_3\,x_4\ldots$ and $y = 0.\,y_1\,y_2\,y_3\,y_4\ldots$, each of them can be paired with an element of interval [0 1]. Given that [0 1] has cardinality c, the union of all points in the unit square can be thought of as a set with cardinality c^2. On the other hand, it is not difficult to show that each pair (x, y) can be matched with a *different* point in set [0 1]. Indeed, given $x = 0.\,x_1\,x_2\,x_3\,x_4\ldots$ and $y = 0.\,y_1\,y_2\,y_3\,y_4\ldots$, their counterpart on this interval can be uniquely determined as $z = 0.\,x_1\,y_1\,x_2\,y_2\ldots$, which implies that

$$c^2 = c \tag{13.12}$$

13.4 Infinite Ordinals and Cardinals

The relationship between cardinals and ordinals changes rather dramatically when it comes to infinite sets. Transfinite cardinals and ordinals represent very different types of numbers, and are subject to different algebraic operations. As noted in the previous section, the infinities that are associated with cardinals have to do with set sizes (as defined by Cantor's pairing metaphor), and for them the operation of addition represents a metaphoric extension of the union of two sets. Thus, an expression like $\aleph_0 + \aleph_0$ needs to be interpreted as the size of a set obtained from the union of two countably infinite sets. The same type of reasoning can be used to derive all other algebraic properties of transfinite cardinals.

For ordinals, on the other hand, operations such as "+" and "·" refer to the *positions* of elements in an ordered set, and have a very different meaning. As a result, transfinite arithmetic with ordinals produces expressions such as $\omega + 1 \neq \omega$ and $1 + \omega \neq \omega + 1$, which have no counterpart in the domain of cardinals (where $\aleph_0 + 1 = 1 + \aleph_0 = \aleph_0$). Nevertheless, it turns out that these two types of numbers are intimately related. To see why this is so, we should recall that each ordinal can be identified with the set of its predecessors. This property allows us to say that ω is *equivalent* to the set of natural numbers, N, and is therefore *the first ordinal that has cardinality* $\aleph_0$.

Once we recognize this connection, it becomes perfectly reasonable to claim that $\aleph_0$ and ω are really two ways of describing the same number. Proceeding in a similar manner, we can define $\aleph_1, \aleph_2, \ldots, \aleph_\omega$ and so on as the *first* ordinal that represents a set of that size. Such a definition provides us with an elegant way to unify two seemingly different concepts of infinity, and allows us to think of cardinals as a "special" subset of ordinals.

It is important to keep in mind that although all cardinals can be thought of as ordinals, the opposite is not true. We can see this very clearly if we observe that ω is the *only* ordinal that is identified with $\aleph_0$, although $\omega, \omega + 1, \omega \cdot n$, ω^n and $^n\omega$ all have the same cardinality (since the sets that define them can be mapped into N in a 1-to-1 manner). This inherent "asymmetry" suggests

13.5. INFINITESIMALS AND HYPERREALS

that ordinals provide a "finer" distinction between different types of infinities, since there are many such numbers between $\aleph_0$ and $\aleph_1$ ($\aleph_1$ is associated with ε_0, which is the first "uncountable" ordinal).[16]

We should also note in this context that each cardinal is actually associated with *two* different ordinals ($\aleph_0$, for example, is associated with ω as well as with 0, which is its index). If we compare these two ordinals, what jumps out at us immediately is the discrepancy in their size. The ordinals that are identified with $\aleph_0, \aleph_1, \aleph_2, \ldots$ represent a hierarchy of ever increasing infinite numbers, while the corresponding indices are small numbers which seem to grow at a much slower pace. Based on this observation, one would expect that the indices could never "catch up", and that this discrepancy holds for *all* cardinals. It turns out, however, that this is *not* the case, since there is actually an infinite ordinal κ which is so large that its cardinality is $\aleph_\kappa$. It is difficult to grasp what this number might look like, but such situations are by no means unusual when it comes to infinite sets.

13.5 Infinitesimals and Hyperreals

Transfinite cardinals and ordinals are not the only infinite numbers that we can define, nor are they the only ones that lend themselves to basic algebraic operations. It turns out that such numbers can be formed in a rather different way, using the notion of *infinitesimals* (which we introduced in Section 12.2). To see how something like that is possible, we first need a more precise understanding of what this term actually means.

One can define the set of infinitesimals by forming a sequence of sets whose n-th term (denoted S_n) represents the set of all numbers a that satisfy $0 < a < 1/n$, and conform to Axioms 1-9 for real numbers. For any finite n, the set S_n obviously contains all the real numbers that belong to interval $(0, 1/n)$, but it remains unclear what happens to this sequence when $n \to \infty$. A natural way to resolve this problem would be to "complete" the iterative process, and assume that there exists a unique set S_∞ whose elements satisfy Axioms 1-9, and are *greater than zero but smaller than any real number*.

We can use a similar approach to define a "representative" element of set S_∞ (the so-called "first" infinitesimal). In order to do that, we should first observe that the number $a_n = 1/n$ has the following four properties:

Property 1. a_n has the form $a_n = 1/n$, where n is an integer. This integer is *larger* than all real numbers on the open interval $(0, n)$.

Property 2. a_n is greater than zero, and smaller than $1/k$ for $k = 1, 2, \ldots, n-1$.

Property 3. For any finite real number r, the product $r \cdot a_n$ is greater than zero and smaller than $r/(n-1)$.

Property 4. a_n satisfies Axioms 1-9.

If we continue to construct such numbers indefinitely, we can assume that there exists a unique number a_∞ such that:

Property 1a. a_∞ has the form $a_\infty = 1/H$, where H is an *integer that is larger than all real numbers* (the symbol H stands for "huge number").

Property 2a. a_∞ is greater than zero, and smaller than all numbers $1/k$ (where k is a natural number).

Property 3a. For any finite real number r, $r \cdot a_\infty$ is greater than zero but smaller than any real number.

Property 4a. a_∞ satisfies Axioms 1-9.

The number a_∞ defined in this manner is known as the *first infinitesimal*, and is denoted by the symbol ∂.[17]

To get a better sense for how infinitesimals differ from real numbers, we should first note that for any two real numbers a and b (where $b > a$), it is always possible to find a third real number c such that $c \cdot a > b$. This property (which is known as the Archimedan Principle) *does not* hold true if a is assumed to be an infinitesimal, since there is no finite number c such that $c \cdot \partial > b$. We can actually make an even stronger statement along these lines, and claim that no algebraic operation can transform an infinitesimal into a real number (which is why these two kinds of numbers are said to be *incommensurable*).

As we noted in Section 12.1.2, real numbers can be formally defined as mathematical objects that satisfy a set of 10 axioms. Infinitesimals satisfy only the first nine, but *not* the Least Upper Bound axiom. To see why this particular axiom is problematic, we must first describe the notion of a *monad*, which was originally introduced by Leibniz.

Perhaps the best way to envision a monad $M(r)$ is to think of it as a "cloud of infinitesimals" that surround the real number r. Such a metaphor makes sense if we observe that adding an infinitesimal to a real number can never produce a *different* real number. Indeed, if we were to assume that $r_1 + \partial = r_2$ (where $r_2 \neq r_1$), then there would exist a real number $q = r_2 - r_1$ that is equal to ∂. This is clearly a contradiction, since ∂ is *smaller* than any real number. As a result, it is fair to say that all numbers of the form $r + \partial$ are *uniquely associated* with the real number r.

Following the line of reasoning outlined above, it is not difficult to show that two monads can *never overlap*. In order to do that, let us assume that the *opposite* is true, and consider two real numbers r_1 and r_2 such that $r_2 > r_1$. If their monads overlapped, there would be a pair of infinitesimals ∂_1 and ∂_2 such that $r_1 + \partial_1 > r_2 - \partial_2$. That, however, is clearly impossible, since it would

13.5. INFINITESIMALS AND HYPERREALS

imply that $\partial_1 + \partial_2 > r_2 - r_1 = q$ (i.e., that a sum of two infinitesimals is larger than a real number).

The fact that monads cannot overlap allows us to prove the following Lemma, which sheds some light on why Axiom 10 is not applicable to infinitesimals.

Lemma 13.1. *Given an arbitrary real number r, the corresponding monad $M(r)$ cannot have a least upper bound.*

Proof. By definition, a least upper bound would be the *smallest number L* that is larger than *any* member of $M(r)$. Let us assume that such a number exists. If L happens to be a *real* number, it will have its own monad $M(L)$, and we know that this monad cannot overlap with $M(r)$. A visual representation of this situation is provided in Fig. 13.10, which indicates that there exists a number $\theta = L - \partial$ which is *larger* than any member of $M(r)$, but *smaller* than L. This is obviously a contradiction, so we must examine other possible scenarios.

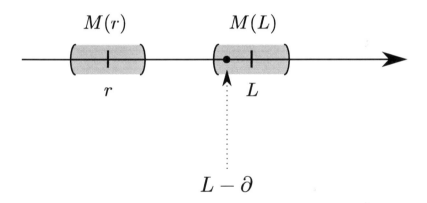

Figure 13.10: Monads $M(r)$ and $M(L)$.

The only plausible alternative is to assume that the least upper bound of $M(r)$ is *not* a real number and has the form $L = r_0 + a\partial$, where r_0 and a are real, and ∂ is an infinitesimal. Such a number would obviously belong to monad $M(r_0)$, which cannot overlap with $M(r)$. Let us now consider the number θ, which is defined as $\theta = r_0 + a\partial - \partial^2$. This number is obviously smaller than L by construction, but it must be *larger* than any member of $M(r)$, since it belongs to monad $M(r_0)$ (as shown in Fig. 13.11). This contradicts our assumption that L is the least upper bound of $M(r)$. **Q.E.D.**

If we add infinitesimals to the number line, its structure becomes considerable more complex. The first thing to observe is that the distribution of real numbers along this line becomes *sparse*, since each one is surrounded by its own monad. "Zooming in" would allow us to differentiate between numbers such as 5, $5+2\partial$ and $5-3\partial$, for example, which are indistinguishable in standard arithmetic. A second "zoom" would draw an ever finer distinction, and would show that a number like $5+2\partial+\partial^2$ is different from $5+2\partial-5\partial^2$. This process could obviously continue indefinitely, uncovering layer after layer of increasingly finer structures, which is why we refer to numbers of the form $x = a_0 + a_1\partial + a_2\partial^2 + \ldots + a_n\partial^n$ as *granulars*. This term was chosen to underscore the resemblance between the nested patterns that we see on the number line and those that are observed when photographs (or fractals, for that matter) are magnified.

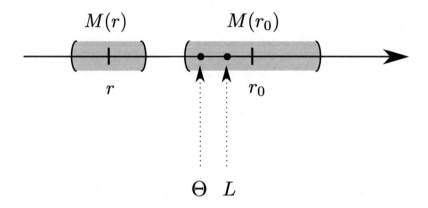

Figure 13.11: Monads $M(r)$ and $M(r_0)$.

The way we defined the first infinitesimal ∂ allows us to introduce infinite numbers that are quite different from transfinite cardinals and ordinals. To see how this can be done, we should first recall that integer $H = \partial^{-1}$ is well defined (this follows from Property 1a), and is larger than all real numbers. Since it is an integer, it has its *successors* $H+1$, $H+2$, $H+3$, ... as well as its *predecessors* $H-1$, $H-2$, $H-3$, all of which are still "huge" (i.e. larger than all real numbers). Since H satisfies the first nine axioms for real numbers, we can also define numbers such as $5H$, H^3 and even extraordinarily large integers such as H^{H^H}.

The collection of all inifintesimals and huge numbers constitutes the set of *hyperreals*. In this set there are numbers such as ∂^H, ∂^{H^H}, $\partial^{H^{H^H}}$ and so on that are infinitely smaller than ∂ itself. To make the notation more manageable, such numbers are usually represented as $\partial^{H(1 \text{ level})}$, $\partial^{H(2 \text{ levels})}$, $\partial^{H(3 \text{ levels})}$ ... $\partial^{H(n \text{ levels})}$ and so on. If we continue this process indefinitely, we arrive at number $\partial^{H(H \text{ levels})}$, which is referred to as "delta super-two", or $^2\partial$. Once such a number is defined, we can easily construct $^2\partial$, $^3\partial$, ... $^n\partial$, ... and eventually postulate the existence of $^H\partial$ as well. This number (which is know

13.5. INFINITESIMALS AND HYPERREALS

as a *superinfinitesimal*) is unimaginably small, but it is by no means the "end of the line" - the recursive construction procedure can continue beyond this point.

Since each infinitesimal satisfies Axioms 1-9, it is guaranteed to have a reciprocal. As a result, we can use the inverses of numbers such as ∂^H and $^H\partial$ to generate multiple layers of "huge" numbers, all of which are infinitely larger than H. It is important to emphasize once again that these infinite "hyperreals" are very different from transfinite cardinals and ordinals - transfinite cardinals arise from Cantor's "pairing method", ordinals are formed by extending the notion of a "successor" in a sequence, while infinite hyperreals are defined as reciprocals of infinitesimals. These differences are so profound that it is fair to say that infinite hyperreals and transfinite numbers represent essentially unrelated conceptions of infinity.

Chapter 14

Notes and References

14.1 Notes for Chapters 12-13

1. George Lakoff and Rafael Nuñez, *Where Mathematics Comes From*, Basic Books, 2000.

2. It is interesting to note that we are actually dealing with two different "layers" of infinity in this case – the set R_∞ has infinitely many members, and each of its elements contains infinitely many decimals.

3. It is important to recognize that each number in this table is rational by construction. It is therefore correct to say that ϕ can be accurately represented as the limit of a sequence of fractions, although it is an irrational number. Table 9.2 suggests, however, that the numbers that define these fractions tend to grow *unboundedly*, so it is of interest to determine how closely ϕ (or any other real number, for that matter) can be approximated if we place certain constrains on the size of their denominators. This is the sort of problem that Diophantine approximation theory is concerned with.

4. Euler eventually managed to solve an entire class of similar problems, which involve sums of the form

$$S = \sum_{n=1}^{\infty} \frac{1}{n^k} \qquad (14.1)$$

where k is an *even* number. In all cases, the results he obtained directly relate to π, and allow us to approximate it using partial sums of fractions.

5. Quoted in: Eli Maor, *To Infinity and Beyond: A Cultural History of the Infinite*, Princeton University Press, 1991.

6. Ibid.

7. Quoted in: Rudy Rucker, *Infinity and the Mind*, Princeton University Press, 2005.

8. For more details on this subject, see: Stewart Shapiro (Ed.), *The Limits of Logic: Higher-Order Logic and the Löwenheim-Skolem Theorem*, Routledge, 1996. Another good resource is: Calixto Badesa, *The Birth of Model Theory: Löwenheim's Theorem in the Frame of the Theory of Relatives*, Princeton University Press, 2004.

9. A thorough discussion of Gödel's theorems and their proof can be found in: Raymond Smullyan, *Gödel's Incompleteness Theorems*, Oxford University Press, 1991 and Ernest Nagel and James Newman, *Gödel's Proof*, New York University Press, 2001. For the original proof, see: Kurt Gödel, *On Formally Undecidable Propositions of Principia Mathematica and Related Systems*, Dover, 1992.

10. Note that the ordinal number of the last element in this sequence doesn't change if we permute its elements. Indeed, if we were to rearrange the original sequence so that $x_1 = 2$, $x_2 = 5$, $x_3 = 7$, $x_4 = 1$, its last element would still be the fourth one.

11. For those who prefer a more formal approach, we should note that the existence of set S_ω is guaranteed by the so-called Axiom of Infinity (which is the fifth axiom in the Zermelo-Fraenkel model).

12. It is interesting to note in this context that Cantor was not the first to recognize that an infinite set can be paired with one of its proper subsets. Several centuries earlier, Galileo observed that there is a 1-to-1 mapping between the set of natural numbers and the set of perfect squares (the correspondence between them is shown in Table 14.1). He recognized that there was something deeply troubling about this result (since the set of natural numbers obviously contains "more" elements), but he made no attempt to resolve the problem. Cantor, on the other hand, embraced this paradox, and used it to define an infinite set as "a set that can be put into 1-to-1 correspondence with a proper subset of itself."

14.1. NOTES FOR CHAPTERS 12-13

$$f(n) = n^2 : \quad 1 \quad 4 \quad 9 \quad 16 \quad 25 \quad \cdots$$
$$\updownarrow \quad \updownarrow \quad \updownarrow \quad \updownarrow \quad \updownarrow$$
$$n : \quad 1 \quad 2 \quad 3 \quad 4 \quad 5 \quad \cdots$$

Table 14.1. Correspondence between natural numbers and perfect squares.

13. Cantor's proof that there is a 1-to-1 mapping between interval [0 1] and the real line suggests that there is a fundamental difference between points on the number line and physical "dots". From a geometric perspective, it would make no sense to say that a line contains as many "dots" as one of its segments. However, such a statement sounds perfectly reasonable if we apply it to *numbers*.

14. This does not mean, of course, that we can multiply 2 by itself $\aleph_0$ times - $2^{\aleph_0}$ is just a *symbol*.

15. Note that the set of real numbers can be partitioned into proper subsets that correspond to *rational* and *irrational* numbers. Since the set of rational numbers has cardinality $\aleph_0$, it follows that the set of irrational numbers must have cardinality c. If this were not the case, we would have

$$\aleph_0 + \aleph_0 = c \tag{14.2}$$

which is impossible (in light of identity (13.9)).

16. The fact that the infinite ordinals that precede ε_0 are no "larger" than the set of natural numbers is surprising and somewhat paradoxical. This set contains numbers such as $\omega + n$, $\omega \cdot n$, ω^n and $^n\omega$ which are unimaginably large, and yet it turns out that each of them can be placed into a 1-to-1 correspondence with N. This is a good example of how our intuition fails us when it comes to infinite sets.

17. Property 2a implies that ∂ is greater than zero but smaller than any real number, which means that $\partial \in S_\infty$. Property 3a implies that multiplying ∂ by a finite number cannot produce a real number, and Property 4a ensures that standard algebraic operations apply to infinitesimals. This last property allows us to define quantities such as 5∂, ∂^m, $1/\partial$ etc., and assume that the usual relations $\partial^n/\partial^m = \partial^{n-m}$ and $\partial^0 = 1$ remain in effect. It also implies that infinitesimals can have different "orders of magnitude", which are proportional to ∂, ∂^2, ∂^3, and so on.

14.2 Further Reading

Number Theory and Infinite Series

1. Konrad Knopp, *Theory and Applications of Infinite Series*, Dover, 1990.

2. George Andrews, *Number Theory*, Dover, 1994.

3. G. H. Hardy and E. M. Wright (revised by Roger Heath-Brown and Joseph Silverman), *An Introduction to the Theory of Numbers*, Oxford University Press, 2008.

4. David Masser, Yuri Nesterenko, Hans Peter Schlickewei, Wolfgang Schmidt, and Michel Waldschmidt, *Diophantine Approximation*, Springer, 2008.

5. Janos Galambos, *Representations of Real Numbers by Infinite Series*, Springer, 2008.

6. Isidore Hirschman, *Infinite Series*, Dover, 2014.

Set Theory

1. Alexander Abian, *The Theory of Sets and Transfinite Arithmetic*, W.B. Saunders Co., 1965.

2. Patrick Suppes, *Axiomatic Set Theory*, Dover, 1972.

3. Martin Zuckerman, *Sets and Transfinite Numbers*, Macmillan, 1974.

4. Thomas Jech, *Set Theory*, Springer, 2006.

5. Akihiro Kanamori, *The Higher Infinite: Large Cardinals in Set Theory from Their Beginnings*, Springer, 2008.

6. Paul Cohen, *Set Theory and the Continuum Hypothesis*, Dover, 2008.

7. John Stillwell, *The Real Numbers: An Introduction to Set Theory and Analysis*, Springer, 2013.

Made in the USA
San Bernardino, CA
19 May 2018